SCIENCE COMMUNICATION IN THEORY AND PRACTICE

Science & Technology Education Library

VOLUME 14

SCOPE

The book series *Science & Technology Education Library* provides a publication forum for scholarship in science and technology education. It aims to publish innovative books which are at the forefront of the field. Monographs as well as collections of papers will be published.

The titles published in this series are listed at the end of this volume.

Science Communication in Theory and Practice

Edited by

SUSAN M. STOCKLMAYER

MICHAEL M. GORE

and

CHRIS BRYANT

Australian National University, Canberra, Australia

KLUWER ACADEMIC PUBLISHERS
DORDRECHT / BOSTON / LONDON

A C.I.P. Ctalogue record for this book is available from the Library of Congress.

ISBN 1-4020-0130-4 (HB)
ISBN 1-4020-0131-2 (PB)

Published by Kluwer Academic Publishers,
P.O. Box 17, 3300 AA Dordrecht, The Netherlands.

Sold and distributed in North, Central and South America
by Kluwer Academic Publishers,
101 Philip Drive, Norwell, MA 02061, U.S.A.

In all other countries, sold and distributed
by Kluwer Academic Publishers,
P.O. Box 322, 3300 AH Dordrecht, The Netherlands.

Printed on acid-free paper

Printed in the Netherlands.

TABLE OF CONTENTS

FOREWORD

Hereafter you will find a first: a comprehensive textbook on the communication of science in theory and practice! Is there a need for such a textbook? Obviously, yes!

Whether you are a scientist, a science journalist, a science teacher, a science museum specialist, a scientific website designer, a science historian, a science entertainer, a member of a scientific society or club, a science tinkerer - there is always a chance that you will be enthusiastic about the job of communicating science, but uncertain about whether you do it well. This book exposes you to the practices of others by reviewing science communication case studies or by inviting you to plunge into the underlying rationale of theoretical approaches to science communication. These are different opportunities for comparing your experience with the practices or reflections of others. The conversation can even be established and furthered with the authors of some of these case studies or of the theories presented in this book by using their website and possibly their email addresses.

This book, like a composite, gathers contributions from experts from most of the above mentioned fields. A lot of practices are discussed and they are among the very best practices according to common professional wisdom in the field. Those practices are also reflected upon in an attempt to be somewhat theoretical. Indeed, the study of how science is communicated is also a science, but it is a social science and as such, has developed theoretical grounds which are solid but differ in their status from theories in natural sciences. Science communication researchers will find in this book concepts and theories to further elaborate on in a field which is moving fast.

Needless to say, the field is complex, requiring more research and more researchers of the highest talent in the future. Among some of questions examined, one can find those concerning the production of scientific knowledge in different contexts, the transposition of the produced knowledge into transferable or reconstructible knowledge, the appropriation mechanisms which depend on the diversity of practices, the balance between individual experiences and representations, between knowing how and knowing about, the use of different genres of discourses and rhetoric, the relation between sciences and between science and other disciplines or domains like technology (the art of solving problems through design), the balance between individual needs (curiosity, health, career) and social needs (as an active member of a manifold of social groups and a society at large) and many more. All of the above represent major questions, the clarification of which has an impact on each of us individually and our place within science-rich societies.

The stakes are high and the challenge of the editors has been well addressed. I have taken a lot of pleasure reading the texts, which offer stimulating inspiration for further practices and reflections. The practice and the future of both science and of science communication are deeply intertwined. This book brings clarity on this matter so essential to evryone today. It is also a call for more responsibility to be taken by scientists in trying to communicate about what they do, what they understand, what they still ignore, and most importantly, what gets them excited.

GOÉRY DELACÔTE
Executive Director,
Exploratorium

ACKNOWLEDGEMENT

The editors wish to acknowledge the hard work and keen eye of Catherine Rayner, a Graduate Student at the National Centre for the Public Awareness of Science at ANU, who assisted in the preparation of the typescript.

SUE STOCKLMAYER, MIKE GORE AND CHRIS BRYANT

INTRODUCTION AND OVERVIEW

It is widely accepted that the importance of the communication of science to the public can be summarised under five headings. They are, not necessarily in the order of importance, economic, utilitarian, democratic, cultural and social.

The economic imperative, regrettably, is today the main driving force towards a better scientifically educated public. No better illustration of this is the joint statement of March 2000 that President Clinton and Prime Minister Blair saw fit to utter, that the human genome is the common property of humanity and not the intellectual property of a few entrepreneurial commercial companies. The concept of intellectual property, however, has fuelled many of the arguments in favour of public awareness of science. This is implicit in the Clinton statement that stresses the attainment of national goals.

The utilitarian argument is closely allied to the economic one. It is the view that the public should be scientifically aware because of the way the community uses science. It is often stated that science in everyday life is invisible, taken for granted. And so it is and–probably–so it should be. It is not necessary in everyday life to know how a magnetron works to run a microwave oven, nor much about electricity to be able to switch the light on and off. It is, however, desirable for the public to keep abreast of the general developments in science becoming aware of new applications, such as the use of DNA fingerprinting in identifying criminals, and feeling comfortable with them.

In a sense, the democratic argument is a subset of the utilitarian argument. The general public is often asked to make decisions about new technologies that could have far reaching effects, both on its own wellbeing and on the rest of the world. Trading in carbon futures in an attempt to halt global warming is an example.

Unfortunately, few politicians or financial experts have any scientific training at all. If the expertise is not present at the highest levels where does it reside? The answer is that it has to reside with the community that, through the exercise of democratic rights, appoints the politicians. It is not just wishful thinking to imagine that, eventually, political will, as expressed by scientifically aware politicians, will have an influence on multinational corporations.

Next, there is the cultural argument. Science is one of the things people do and, like all of the things that people do, it can be done at the highest or the lowest level. As Stephen J. Gould has remarked, the best science is like high art, worth appreciating for its own sake and not necessarily because it brings an immediate benefit in material terms to the beholder. The elegant simplicity of Nobel laureate Niko Tinbergen's studies of digger wasps and herring gulls had no immediate 'use', but for the reader the sheer pleasure of comprehending something of the life of these common animals transcends the need for 'usefulness'. Although a lesser scientist might have achieved the same results, Tinbergen's insight, the creativity of his hypotheses, the elegance of his experiments and the simplicity of his writing combined to produce work that takes science to the highest levels of human endeavour. Mathematicians experience the same *frisson* of pleasure when contemplating an elegant solution to a problem even though a 'quick and dirty' answer might do the job.

Finally, there is the social argument. As science permeates all levels of human activity then an awareness of the basis of science and the issues surrounding it will serve to enhance social cohesion. Much of science has relevance for all cultures and becomes a shared tradition. Unfortunately, it cannot be denied that, as Gregory and Miller (1998) point out

> There is no doubt that many in the scientific community who want to further the public understanding of science are really concerned with increasing the public's appreciation of science, with a view to enhancing the status of themselves and their colleagues. (p9)

Scientists are very likely to think in terms of the way they perceive that scientific communication might benefit their own work. A common feeling amongst scientists, as soon as they emerge from the self-delusion of believing that people will not understand what they do, is

> if only people knew how exciting my science is they'd give us more money

In spite of much evidence to the contrary, they rarely think

> if people knew what we are really up to, they might stop us

Scientists are no more nor no less altruistic than non-scientists. Even men like Humphrey Davy, Michael Faraday and Thomas Huxley, all of whom were outstanding science communicators, were driven by the twin imperatives of the science itself and the need to make a living. Good communication was therefore essential.

As far as the community is concerned, science is invisible until such time as it has a need for it. It is the task of the science communicator to demonstrate to the community that it has such a need. When the need is recognised the rate at which non-scientific members of the community are capable of assimilating scientific ideas is often astonishing.

There have been many surveys that claim that the public would prefer to read about science ahead of sport in the newspapers. They are often cited as a validation for the science communicator and one that should convince editors to include much more science in their papers. The editors, who presumably know what sells newspapers, remain unconvinced, and so does Professor Doherty (see Epilogue).

Yet it is an inescapable fact that several surveys have come up with the same conclusion. The reading public likes to read about science and places it high on their list of priorities. Editorial policy, however, is informed by circulation. It is clear that editors are comfortable with the view that, while the reading population might like to read about science it doesn't want to read about it too much.

In understanding the community's attitudes to science, it is necessary to distinguish between 'knowledge' of science and 'affect' or mental disposition towards science. Herein lies the dichotomy between the public understanding of science (which rejoices in the revolting short form PUS; as someone remarked, only scientists would come up with an acronym like that) and public awareness of science.

It is good to be sceptical of studies that indicate that the public has a poor grasp of science. Many of these studies contain the seeds of self-fulfilling prophecy. For example, Durant Evans & Thomas (1989) describe work in which general scientific knowledge questions are asked of a randomly selected sample of people. This necessarily constructs an image of a public that is deficient in its understanding of science. Only if everyone answered all questions correctly could this deficit model be challenged. Other studies may have other flaws; if the public is asked what it wants to read about in newspapers, it may give 'science' as the socially desirable answer, or confuse science with medical breakthroughs. If the community is asked whether it is proud of the achievements of its scientists it might answer 'yes' without having a clear idea of exactly why. We can all be proud of the Dohertys and the Floreys without necessarily understanding what they did to win the Nobel Prize.

Only when the public have a specific interest will they turn to the science pages. When they are personally engaged, they will read voraciously and be capable of mastering difficult material with ease. It is the task of the science communicator to increase the 'need to know' and nurture it.

This seems so obvious that it is a surprise when scientists fail to grasp this very simple idea. In the face of declining enrolments in science at secondary and tertiary levels in the western world, and of the obvious disenchantment of young women with science, there is still great resistance amongst scientists to making their work accessible.

Research Councils now include the concept of 'duty' in their instructions to applicants. The Particle Physics and Astronomy Research Council in Britain, for example, states that

> We believe that all those engaged in publicly funded research have a duty to explain their work to the general public.

This imperative generates both anger and anxiety among scientists when confronted with it. One of us (CB), in an earlier incarnation, recalls the intense irritation he felt when he had to spend valuable research time on both grant applications and communication with the public. These are feelings shared by many scientists today. We frequently run short, 3-day workshops on science communication for scientists. We have noticed marked variations in response to our opening question 'why bother with science communication?'

The more senior scientists tend to express resentment. They are in denial—their position is summarised by this statement:

people don't need to know about science. They should just let us get on with the job.

When we argue strongly that the community needs to become more scientifically there is a subtle shift of position:

I'm a scientist. It's what I do best. Let other people do the communicating.

Their point of view is easily understood. They grew up in a world of certainties, entrained into science at an early age, secure in the knowledge that there were facts somewhere 'out there' to be discovered and that, once discovered, these facts would immediately be accepted by an admiring public. They were never expected to communicate and may even have gone into science because they were poor communicators.

The ideas that science is culturally dependent, that knowledge is constructed, threaten their mastery of their discipline. They feel uncomfortable, unhappy and yearn for the certainties and security of their laboratories and their white coats. From this position it is a short step to agreement that while science communication might be a good thing it is not for them because

people wouldn't understand what I'm doing.

The sub-text is that their science is so complex and they have invested so many years in acquiring their expert status that they find it difficult to imagine that it can be understood by someone who has not put in the same effort. It threatens both their standing with their colleagues and their self-respect.

This idea that their science might be accessible to a lay public is abhorrent to many western scientists who seek certainty and absolute truth. Theirs is a view akin to those of the mediaeval guilds that protected their knowledge by allowing only initiates into full understanding of the one truth. Knowledge was power and was not to be let go; its communication was only permitted between initiates. It often required special language and signs; and it was not for the general public.

Graduate students and newly fledged scientists do not hold these views so rigidly. They are still only on the verge of disenchantment and have largely retained their enthusiasm for their subject and even some proselytising zeal. It is imperative that this be nurtured.

In summary, increasing the public understanding of science is a worthwhile endeavour that creates an intelligent, informed and skilled group within the community. Such a group is an extremely valuable resource for the community. Increasing public *awareness* of science, however, is a longer term project, but one that, if successful, can contribute enormously to social well-being as it creates a community that is confident in its possession of scientific ideas and is comfortable about raising children to have the same confidence.

It is the intent of *Science Communication in Theory and Practice* to range far across the whole field of science communication. It gives both a theoretical basis for the newly emerging discipline and a series of personal histories, the authors of which are effectively saying 'this is what *we* did' and in all cases, they did it very successfully. They are offering their experience but none of them is prescriptively saying 'this is what you should do'. What they hope is that the prospective science communicator will shop amongst them and use whatever is fitting for his or her circumstances.

The structure of the book may need some explanation. It falls naturally into four parts, and an Epilogue. Part 1 consists of five chapters that are concerned with the disciplinary basis of science communication. Susan Stocklmayer explores images of science together with the implications they have for the effective communication of science. She dissects gender issues and draws attention to the 'masculinity' of science–the majority of practitioners are male, and thus are poor role models for girls. She points to the gender bias of texts, and dismisses biological factors as the prime determinant. She describes a constructivist model for learning and considers the consequences this has for communication. Glen Aikenhead sees science as culture, and the interaction of this culture with the 'culture of a public immersed in their everyday lives' is the essence of science communication. Science communication is 'crossing the border' and, when those borders are not crossed successfully confusion reigns. One of the editors still recalls with discomfort the debilitating gastric overload when the politeness, acquired as a child, of 'finishing everything on your plate' clashed with the politeness and generosity of his Indian hosts, to whom an empty plate meant 'please may I have some more'. Overloads are often experienced during the process of communicating science and, because most people are polite and have no wish to hurt feelings, the mores of the two cultures are not explored. Science communication permits an unthreatening exploration of the two cultures.

Jon Turney examines the depiction of science in literature, and science communication as a genre in its own right, and reflects on what qualities are present in good popular science writing. As the genre draws its inspiration from such a wide range of human endeavour he wonders whether it will ever be possible to categorise it. Lawrence Prelli sees science as discourse and discusses the implications of rhetorical understanding of argumentation in science, and the implications for scientific literacy. He has some trenchant comments on the science literacy movements –

> if the public cannot grasp these essentials (a majority of Americans in one survey believe that Israel is an Arab nation) why should we expect that science literacy campaigns can elevate their comprehension of the complex principles of science?

These prophetic words have recently been echoed by an influential report of the House of Lords who wish to move the game from 'public understanding' to 'awareness' of science.

Richard Eckersley takes a *fin de siécle* look at science and sees harbingers for the loss of confidence in modern science at the beginning of the last century. He finds science at a threshold of opportunity provided by science communication. It remains to be seen whether the opportunity will be seized; already the first skirmishes - mad cow disease and genetically modified foods, for example - seem to be lost.

Part 2 is concerned with research into the learning of science in informal situations and it is still a new field. In fact, Sless and Shrensky draw a powerful comparison between science communication and rain making and suggest that the basis for believing that it is effective is weak or none-existent. This iconoclastic view is challenged by Léonie Rennie. Rennie is one of the few with substantial qualifications in the area; in particular, her recent research has focused on the impact of science centres on the community. She points out that asking whether people learn science on a visit to a science centre is to ask the wrong question. It is far better to ask whether visits to science centres contribute to improved long term relationships with science. John Gilbert strives towards an understanding of the competing roles of education and entertainment in science communication, illustrating it with original research. Susan Stocklmayer then summarises the 'understanding' versus 'awareness' debate, comparing and contrasting the formal and informal learning of a science with a model based on the constructivist view.

Part 3 comprises three chapters dealing with aspects of the media and their role in science communication. Peter Spinks, an eminent journalist with much experience on major papers in the UK and Australia, distils his experience into a manual for scientists on working in and with the print media. Deborah Cohen, of the British Broadcasting Corporation, describes the art of making radio science programs, while Ian Allen, of the Australian Broadcasting Corporation does the same for television. Allen also discusses the presentation of good science on the world wide web and contrasts this new art with the long established one of making television science shows.

The practical aspects of science communication, in all their manifestations form the heart of the next five chapters and make up part 4. Peter Briggs, executive officer of the British Association for the Advancement of science outlines the history of the BA and its role in encouraging the formation of other, similar Associations in different parts of the world. Science festivals are now common throughout the world. Once they were few and far between and the Edinburgh Science Festival is, if not the oldest, certainly one of the first two or three. Simon Gage, the present Director, outlines the forces that gave rise to the Festival and its subsequent history. This chapter is a detailed prescription for mounting such an event and is an invaluable primer for those contemplating it.

Michael Gore and Chris Bryant examine the different sides of a symbiosis, of the coming together of a science centre and a university science faculty. Gore outlines the genesis of Questacon, the National Science and Technology Centre of Australia and along the way offers many interesting insights into such things as exhibit design and science theatre. Bryant, from the Australian National University, is concerned with the last 12 years of fruitful collaboration with Questacon and the growth of a graduate program in science communication, probably the first in the world. He describes how the collaboration finally brought forth a flourishing university centre focusing on the public awareness of science.

A very different model is presented by Vivian Altmann, Modesto Tamez and Dennis Bartels. They base their program on community needs. Their approach employs hands on exploration as a tool for reinforcing social values and affirming the social worth of the individual. They remind us that we should not forget other places that do not need large buildings, budgets and staff to achieve effective science communication.

The Epilogue, by Nobel Laureate Peter Doherty describes a personal odyssey in science communication. His is essentially an account of his overnight transition from eminent scientist to public media figure. It is the story of what he found to be important. Things which might appear self evident to a member of the public - 'use language like a normal human being' - are not necessarily so to scientists used to conversing in a sort of scientific 'shortspeak'. Doherty's insights during his transformation should bring comfort to many lesser scientists uncomfortable with their perceived failure to make themselves understood

This is a book, then, that can be read on several levels. It can be read as a theoretical basis for science communication; as a practical guide for would be science communicators; or as a contemplative projection of what culture of science might be like in the 21st Century if the enterprise of communication is successful.

National Centre for the Public Awareness of Science
Australian National University, Canberra ACT 0200 Australia

REFERENCES

Durant, J. R., Evans, G. A. & Thomas, G. P. (1989). The public understanding of science. *Nature (Lond), 340,* 11-14

Gregory, J. & Miller, S. (1998). *Science in Public: Communication, Culture and Credibility.* Plenum Trade, New York and London.

SECTION 1

TOWARDS A THEORETICAL BASIS FOR SCIENCE COMMUNICATION

SUSAN M. STOCKLMAYER

1. THE BACKGROUND TO EFFECTIVE SCIENCE COMMUNICATION WITH THE PUBLIC

1. INTRODUCTION

To be effective with any audience, communication must be an interactive process. As Sless and Shrensky show in Chapter 6, science communicators who think only of the message and not of the 'audience' are likely to fail. Communication is essentially as much a matter of listening as it is of talking and, to be effective, each party must have some understanding of the other. In this chapter, I shall review what we know about the ways in which the general public views science and scientists and I shall consider some impediments to understanding which, if overlooked, may prevent effective scientific communication.

There is no doubt that communicating science is difficult. Were this not so, there would be no need for any of the chapters in this book. Indeed, there would be no one to write them! All of the authors are in the business of trying to make science more understandable. This chapter is about the public perceptions of science, and so we must immediately make some assumptions about the people we are addressing.

Let us assume, for the moment, that the 'public' is a Western group of people of various socio-economic and cultural backgrounds. This public is male and female, with all the complexities of personality, age and experience now embraced by the term 'gender'. The public has, on the whole, some rudimentary education in science. We know immediately that the form and extent of this education is country-dependent. Some (often female) have had some biology lessons but little or no physics and chemistry. Many remember little of their school science. Most have little knowledge of the earth sciences and most are increasingly conscious of ethical, ecological and environmental issues. The structure of this public looks immensely complex and variable.

Why, then, does world-wide research in the Western world - and in many Asian countries formerly occupied by colonial powers - keep finding the same overall picture? For the past twenty years at least, research coming from the area of science education has revealed a public that is fearful, mistrustful and ignorant of simple scientific principles. Why has education failed to address these problems and what should science communicators know in order to be more effective?

The research spanning two decades is immense. It encompasses the ways in which people learn, how and why people change their ideas, the nature of schooling itself. The theory of 'constructivism' has illuminated some of the reasons why people choose not to pursue science and fail to remember what they have learned - but that is not the whole answer. We know, for example, that part of the problem is

3

S.M. Stocklmayer et al. (eds.), Science Communication in Theory and Practice, 3–22.

that women relate to science less well than men and that science careers are less common amongst disadvantaged groups. It is important also to recognise that science is not "common sense" and that misconceptions abound. This chapter will review only those areas of research which relate directly to the informal learning of science. It is not possible here to address the problems of curricular structure and the nature of schooling. These have, in any case, been addressed at length elsewhere (see, for example, Cobern, 1998). To begin, let us examine the image of science and scientists as revealed by countless participants in the 'Draw-a-Scientist' test.

2. THE POPULAR IMAGE OF SCIENCE

The 'Draw-a-Scientist' test was devised by Chambers (1983) to investigate how children imagined a scientist. Not surprisingly, the most common image is of a middle-aged white male wearing glasses and a white coat. He frequently has facial hair and is often bald on top. He is surrounded by symbols of science that are usually chemistry apparatus - test tubes, flasks and so on. Since the early 1980s the test has been repeated all round the world for various purposes and with varying degrees of criticism as to its validity and meaning (for example: Butler Kahle, 1987; Finson, Beaver and Cramond, 1995; Lannes, Flavoni & De Meis, 1998; Jarvis, 1996; McAdam, 1990; Schibeci, 1986). The test is, according to many critics, flawed in that it encourages participants to perpetuate a ubiquitous media image of scientists - either the stereotypical mad professor of the movies or the friendly adviser who is advertising the most suitable washing powder. There is, without doubt, some substance to this criticism and if all were well elsewhere in the world of science communication we might ignore this phenomenon.

There is a darker side to this image, however, which we need to note. Very often, the drawings depict eccentricity bordering on madness, with an indication of evil intent. The drawings produced by female senior high school students (Figure 1) illustrate this inherent lack of responsibility towards humanity, and attendant cruelty to animals. It is not just *scientists*, therefore, who are depicted in this way – it is *science itself*. Whatever its origins – and one might conjecture that the stereotype goes back all the way to Dr Frankenstein and beyond (Turney, 1998 - it illustrates the underlying beliefs of the general public about who scientists are and what they are about. They are white middle-to-upper-class males, middle-aged, socially inept, at best eccentric and at worst downright evil. The nutty professor images portrayed by so many television science presenters merely confirm what everyone believes.

Can the 'Draw-a-Scientist' test be taken too far, and too much concluded from its results? Let us examine in some detail what it says about science. First and most obvious, science is seen as masculine. This masculine nature of all science, especially the physical sciences, has been much documented but is still hotly debated by many within science and outside it. As a 'seeking after truth', say the critics, how can science have any sort of gendered identity? Yet, as Fox Keller (1985, p.55) explains, members of the Royal Society of the 17th Century which formed the foundation of Western Science explicitly stated that their science was

Figure 1.   Drawings of scientists by 17 year-old female science students in Australia

'masculine and durable', seeking to capture and control a 'female' Nature in ways not hitherto sought. Their science was founded on virtues of objectivity, strength and rigor. It therefore sought truth in a particular 'value-free' way, asking particular questions which were decided and defined by its practitioners. Other, perhaps equally legitimate questions, were never raised because they were of no importance to the scientists themselves.

It is easy to see that the history of science is male-dominated. For all sorts of reasons, few women participated in its endeavours. Passing quickly to the end of the 20th Century, however, we find superficially a very different picture. Western Science is now perceived by many – including large numbers of university students - to offer equal opportunities to women and men and equal voices in the world of science.

> Things have changed since the beginning of science and if too much emphasis is put on encouraging females to take part in science, it could end up by disadvantaging males. (Female Australian science student, 1998)

> Science has not been equally accessible for women until recently... Now there is no difference. Women have the same opportunities as men and should take full advantage of them. I believe there are only men in the top positions because women haven't had enough time to occupy those places. (Male Australian science student, 1998)

Has the picture really changed that much? Kelly (1985) outlined four senses in which science might be described as masculine. The first was in terms of numbers–those who practise it are mainly male. At the end of the 20th century, the numbers have certainly changed in many countries at the school and undergraduate levels. A closer look at the distribution of male and female participants, however, indicates that the biggest increase in female students has occurred in the biological and environmental sciences with the physical and earth sciences lagging far behind. For example, in Australia at least, female participation in undergraduate engineering appears to have reached a plateau at 14%. As one goes higher up the scale of all science careers, the proportions alter dramatically in favour of males.

This bias is less true, however, of many European and Asian countries in which women reportedly form a much higher percentage of the scientific workforce. So perhaps the relative numbers are a cultural rather than a scientific issue, requiring specific initiatives in English-speaking countries to coax more women into taking up long-term careers? This may be true. We need, however, to look at how 'science' is described in some of these countries and what work is defined as 'scientific' before reaching a conclusion (see, for example, Ancog, 1998; Hermawati, 1998.) There is evidence also that despite improved participation rates, women in European science still struggle for equality of opportunity.

Kelly's second area of masculinity was in the packaging of science. She quoted science curricula, textbooks, applications and so on as coming from a male world and describing a male world. By 2000, much had been done to address this issue. Yet an examination of current popular physics texts reveals an overwhelming bias in favour of masculine examples. Those few that feature women are often stereotypical. The textbooks are still, in graphics and language, presenting to a male audience.

The language of science is an area that has attracted considerable criticism. Jargon is endemic in all disciplines but the language of science is *itself* inherently masculine. There has been a move in recent years to re-write texts to be more gender inclusive, including more examples of a 'gender neutral' type. Research into this issue indicates, however, that so-called gender neutral nouns ('an athlete', 'a builder', 'a cyclist') are assumed by female readers of science texts to be male. From my own observation, when a group of 14-year old girls were given a physics problem involving a bird, it evoked responses from all 30 students which referred to the bird as 'he'. Reasons for this are not well understood – but truly gender inclusive terms have been shown to be 'you', 'we', or examples of particular people by name, irrespective of their sex.

Kelly's third area of masculinity was that of practice. Those who instruct students, from school through to postgraduate study, set benchmarks of 'good' and 'bad' science, act as role models and define what science expects of its practitioners:

> Do they praise the loudest 'pops' when children are making hydrogen or the most beautiful soap bubbles? These seemingly insignificant choices set the tone of the lessons and influence the image of science presented to the class as harmful or caring. (Small, 1984, p. 30)

Last, Kelly quoted biological factors. Many of these have largely now been refuted. There is still debate, however, about the differences inherent in male and female brains (e.g. Moir and Moir, 1998). What these differences are is not yet well understood, but the evidence from the world of science is clear. Science is, at present, unappealing to most women and to many men. Its masculine character has developed as a social construction reflecting a patriarchal society. There are many examples, historically and at the present time, of overtly male practices and discrimination in science. The more subtle question of the nature of science itself has, however, also been explored, by feminist scholars such as Belenky, Clinchy, Goldberger and Tarule (1986) who describe 'women's ways of knowing' quite different from those of men. Women, according to these scholars, prefer amongst other things to consider problems holistically. They value intuition and prefer non-hierarchical interactions.

This presents a problem for science if it is still to be regarded as intrinsically objective, abstract and value-free. The sad result will be that it will always remain the domain of a few men of a particular type. The importance of factors such as intuition will continue to be denied, even though it is well established that, in the actual practising world of science, researchers frequently follow their instincts. Indeed, Fensham (1993) reports that a series of British television interviews over seventeen years with Nobel scientists revealed that although six laureates denied the existence of scientific intuition and five were doubtful, the remaining seventy two of those interviewed 'readily acknowledged its existence' (p.16). Fensham goes on to comment that 'the eleven (denials and doubters) then went on to refer to experiences of the same kind as the seventy-two'.

If science accepts that there is room for other views, room to ask other questions and room to accommodate other understandings, then it can grow into a truly universal discipline which accommodates most women as well as many men who have rejected science. The philosopher, Sandra Harding, in her book *Whose Science, Whose Knowledge?* (1991) argues for a more diverse science which embraces other world views:

> Thinking from women's lives provides crucial resources for the invention of sciences for the many to replace sciences that are often only for the elite few. Without such new sciences, privileged groups remain deeply ignorant of important regularities and underlying causal tendencies in nature and social relations and of their own location in the social and natural world. Without such sciences, the majority of the world's peoples remain deprived of knowledge that could enable them to gain democratic control over the conditions of their lives. (p. 312)

Parker (1996, p.17) defines this gender-inclusive science as having the following characteristics:

> an *holistic* as well as an atomistic approach
> *order* as well as law
> *mutual respect* and interaction as well as domination
> *a non-hierarchical continuum of difference* as well as a dichotomy and polarisation
> *involvement* as well as detachment
> *understanding* as well as predicting
> *empowerment* through understanding as well as power to manipulate
> *broadly defined* as well as highly specialised scientific knowledge
> *scientific knowledge contextualised in history and in contemporary society* as well as a historical and decontextualised scientific knowledge

3. CONSTRUCTIVISM

The requirement that science be placed in context is a fundamental tenet of constructivist ideas about learning in general. Constructivist approaches to teaching and learning have transformed some classrooms, requiring of the teacher a recognition that every student will build upon an existing knowledge framework and accept or reject new ideas in the context of what is already known. This presents problems for the formal classroom, in that the ideal environment demands individual attention to personal learning styles, personal previous experience, beliefs, and so on. In the informal context, it becomes even more problematic in that there is usually little opportunity to discover very much about the background of the 'audience'. Science communicators have to take a great deal for granted and make sensible guesses. There are, nevertheless, some basic assumptions which can assist effective communication. They may seem like common sense - yet it is astonishing how often these basic principles are ignored.

First, there is the need for relevance. According to constructivist theory, no learning can occur if there is no foundation on which to 'pin' it. In other words, to remember something means having some sort of memory hook.

> Learning outcomes depend not only on the learning environment but also on the knowledge of the learner… Construction of a meaning is influenced to a large extent by our existing knowledge. (Driver and Bell, 1986, p. 453)

In science communication, this hook is likely to be a reference to an already existing piece of knowledge or experience – that is, to a phenomenon coming from the real world of the audience. The emphasis is on *building* – not factual 'telling'.

> Explaining a problem will not lead to understanding unless the learning has an internal scheme that maps onto what a person is hearing. Learning is the product of self-organisation and reorganisation (Yager, 1991, p. 55)

> Teaching by telling is an ineffective mode of instruction for most students. Students must be intellectually active to develop a functional understanding. (McDermott, 1993, p. 161)

> Students need to be given opportunities to make sense of what is learned by negotiating meaning; comparing what is known to new experiences, and resolving discrepancies between what is known and what seems to be implied by new experiences. (Tobin, 1990)

The real essence of constructivism is participant learning - something not always available to the science communicator. Nevertheless, some important aspects suggested by Yager (1991, p.56) may have application to informal science communication. These are:

i.   Using open-ended questions
ii.   Encouraging participants to suggest causes for particular events and predictions of consequences
iii. Encouraging the testing of the participants' own ideas, e.g. answering their guesses
iv.   Encouraging participants to challenge conceptions and ideas
v.   Using cooperative strategies that emphasise collaboration, respect individuality and use division of labour
vi.   Encouraging adequate time for reflection and analysis, respecting all ideas
vii. Encouraging self-analysis, collection of real evidence to support ideas, and re-formulation in the light of new experiences
viii. Encouraging the use of alternative sources of information, both in written materials and in the use of experts

Yager goes on to identify characteristics which distinguish a constructivist environment. These include emphasis on local interest and local resources. Participants are actively involved in seeking information to solve real-life problems. The impact of science on every individual is emphasised, especially the importance of science and technology for the future. Issues of science also are a part of this environment, with participants assuming a citizenship role in attempting to resolve them. Last, the 'special' quality of science skills and processes is de-emphasised. You do not have to be 'special' to participate fully in science.

Extending right across all this attention to individual needs and experience is the notion that everyone learns in a different way. There have been several different descriptions of learning styles: I shall quote just one here as an example. Honey and Mumford (1986) identified four kinds of learning style:

> *Activists* 'learn by doing. They are 'open-minded... enthusiastic about new experiences, and tackle problems by brainstorming... They are bored by implementation and longer term consolidation... They are gregarious'.
> *Reflectors* 'like to stand back, to ponder experiences... They collect data... and chew it over thoroughly... They postpone reaching definitive conclusions as long as possible... They enjoy observing other people in action, listen to others and adopt a low profile.'
> *Theorists* 'adapt and integrate observations into complex but logically sound theories. They tend to be perfectionists... like to analyse and synthesise. They are keen on basic assumptions, principles, theories and models...and tend to be detached and objective.'
> *Pragmatists* 'are keen on trying out ideas to see if they work in practice. They positively search out new ideas, get on with things and act quickly and confidently. They are impatient with open-ended discussions.'

There are, in any population, for each person, combinations of these styles. Everyone has a little of each, but one style usually dominates. For any communicator, there is an immediate problem of providing for these various styles of learning while attempting to convey ideas through a personal preferred communicating style. If you are not to lose over half your audience, however, it is extremely important to recognise that not everyone shares your own preferred mode of learning and, within your communication, build opportunities for people to learn in different ways.

Learning styles have been extended in the context of interactive science centres to encompass the theory of multiple intelligences (Gardner, 1993). According to this theory, there are seven basic intelligences which have been detected and which need to be addressed by, say, an interactive exhibit if it is to appeal to as wide an audience as possible.

These are:

i.   Linguistic intelligence.
ii.   Logical-mathematical intelligence. This incorporates scientific ability.
iii. Spatial intelligence
iv.   Musical intelligence
v.   Bodily-kinesthetic intelligence. This is the ability to use the body to solve problems, or to fashion products
vi.   Interpersonal intelligence – the ability to understand and work with others
vii   Intrapersonal intelligence – the ability to form an accurate model of oneself and use it to operate effectively (pp. 8-9)

These intelligences, according to Gardner, are distinct. For any individual, one type will have more appeal than another depending on the degree to which that particular intelligence is developed. For effective learning to occur, relevant experiences embodying that intelligence must be provided for. Clearly, interactive exhibits offer a special opportunity to cater for multiple intelligences. Few, however, actually do in practice.

The theory of constructivism embodies the essence of all successful communication - listening to and understanding the audience. Whether that audience be physically present or at a distance (as is the case for journalists and the designers of interactive multimedia), ignoring the nature of the audience and how they think will result in failure to communicate. *Everything* about the ways in which science concepts are presented taps into constructivist theory: problems with relevance, with style, with language.

Another common difficulty with communicating science is the possibility of the 'wrong idea' being conveyed, leaving the audience more puzzled than before. Over the past two decades, misconceptions have been re-named alternative (or sometimes 'naïve') conceptions – because so many of them originate well before any formal schooling in science.

4. ALTERNATIVE CONCEPTIONS

The difficulties and nature of alternative conceptions in science have been identified and researched over many years – indeed, even early scientists such as Michael Faraday were awake to the problem. Faraday commented about the terminology of electricity:

> As long as the terms current and electrodynamic are used to express those relations of the electric forces in which progression of either fluids or effects are supposed to occur... so long will the idea of velocity be associated with them. (quoted in Stocklmayer and Treagust, 1994, p. 141)

Stocklmayer and Treagust identified the 'fluid' language of electricity as one of the major factors causing misconceptions as to its nature.

The notion that children form ideas of their own about many scientific phenomena was first described formally in seminal work by Osborne and Gilbert (1980), which led to the coining of the term 'children's science'. They investigated children's understandings of various physical phenomena and identified alternative conceptions about mechanics and electricity. This work was followed by wide-ranging research into alternative conceptions in physics, chemistry, biology and the earth sciences. In every discipline, and in many different countries, common ideas were identified. Some aspects of some of these are sketched briefly below.

4.1. Alternative Conceptions in Physics

Notable alternative conceptions in physics have been detected in the areas of Newtonian mechanics, energy, light, density, buoyancy, electricity and so on. Many of these are described in *Children's Iideas in Science* (Driver, Guesne and Tiberghien, 1985). I have selected two areas to give a flavour of the findings: these are alternative conceptions about energy and electricity.

4.1.1. Energy

'Energy is an abstract quantity, invented to aid humanity in the investigation of nature. As a result it is almost impossible to store an abstraction.' (Warren, 1986, quoted in Beynon, 1990, p. 315). The concept of energy exists only as it is defined, a fact that was recognised by Feynman (1989) when he said that 'we have no knowledge of what energy is.' It should not be surprising, therefore, that students 'map' lay meanings of energy onto scientific meanings - and that those members of the public who have had no science education understand energy only in a common-sense way. Concepts such as potential and kinetic energy, energy transfer and conservation become very difficult to communicate. For example, as Beynon explains (p. 315), the common reference of ground level for zero potential energy leads many to assume that a rock at rest on the ground has no energy at all. 'Storing' energy as the rock is raised up is a meaningless image for most people. Understanding how energy is used by humans is important for everyday life - after all, we pay for it! - yet this aspect also causes problems for learners (Bliss and Ogborn, 1985; Kirkwood and Carr, 1989).

Warren (1986, p.155) states that you can only understand 'energy' if you understand 'work' and that this concept depends on an understanding of 'force'. He is, of course, speaking as a physicist. Yet the three terms are used every day in the world outside physics. It is this very fact that causes alternative conceptions to be formed at an early age. Boyes and Stanisstreet (1990) identify a number of alternative conceptions about the Law of Conservation of Energy which includes, for 50% of students up to the age of 16, the notion that this law is 'a statutory law prohibiting abuse of the countryside and wildlife' (p.54). The authors attribute this confusion to the 'interchangeability of vocabulary between the scientific and everyday domains'.

When 'heat' is brought into the picture there is even more confusion (e.g. Linse, 1990, Rogan, 1988). Understanding the laws of thermodynamics is very difficult (Solomon, 1982; Ross, 1988) and concepts such as heat conduction require learners to regard heat as, in some sense, a material entity (Clough and Driver, 1985). Linse (1990, p. 576) states that

> heat is a notion with its own confusions in the minds of pupils....Heat, as a form of energy, is made, is generated, is set free, gets lost, and is contained in objects.

Linse concludes that

> natural thinking about energy is characterised by its strongly anthropomorphic, material... causal and non-mathematical character.

His findings were echoed by later work by Trumper and Gorsky (1993). The idea that energy is something material, like a fuel, has been addressed also by Duit (1987) and Ogborn (1986).

Small wonder that any attempts to communicate ideas about energy are likely to run into trouble. Viglietta (1990) quotes Anna, an Italian 15-year old who exemplifies the feelings that the entire energy concept engenders in students:

> I wish I could know everything about energy because you may start a discussion and get trapped into an awkward question which is not a nice feeling.

4.1.2. Electricity

There are many alternative conceptions about static and dynamic electricity. As an example, I shall quote some of the findings about simple DC circuits that have come to light in the past two decades. Solomon, Black, Oldham and Stuart (1985) identified ideas about electricity in quite young children that originated from 'a *common culture* of knowing, saying and emotive reactions which is available to all these children', enabling them to draw on a common 'social stock of life-world knowledge' (p.286). These ideas bore little relation to the physicists' notion of electricity, but compared it to a fire, a river or a dangerous animal.

As students progress with formal schooling, misconceptions are introduced about DC circuitry which lead to wrong predictions of circuit behaviour. For example, Shipstone (1984) detected a hierarchy of understanding for high school students which begins with a belief that the current leaves the battery at both terminals and is used up within the circuit elements (Model 1). Osborne and Gilbert (1980) refer to this as the 'clashing currents' model. Later, students may come to believe that current flows in one direction around the circuit, becoming gradually weakened as it goes so that later components receive less. Lamps furthest along the circuit will be least bright. (Model 2). Many students retain this idea long past the end of schooling. The last alternative conception (Model 3) which Shipstone (1984) identified was a refinement of the second: students think that current is shared between the components in a circuit. If lamps are identical, current will be shared equally but not conserved.

Research has shown that this pattern of modelling of direct current persists in a large number of countries with thousands of students of different ages (e.g., Campbell and Krockover, 1992; Closset, 1983; Duit, 1985; Dupin and Johsua, 1987; Eylon, and Ganiel, 1990; Fredette and Lochhead,1980; Maloney, 1986; Millar and King, 1993; Saxena, 1992; Stocklmayer, Zadnik and Treagust, 1993). Duit (1985) discussed the words 'current' and 'voltage' because these words occur in everyday language, as well as in the classroom. In general, Duit identified three sources of learning difficulties associated with language (p. 212): first, language provides images which convey meaning contrary to scientifically established facts (eg. 'the sun rises'). Second, the structure of language may mislead: voltage is a concept which has a 'quality' in everyday terms but which is used by physicists in a relational manner. Third, concept names occur in everyday language in a manner different from their scientific meaning. The word 'current' in English is equivalent in German to *Strom* (stream); both have their origins in early fluid notions of electricity but mislead students of physics today. The German word for voltage is *Spannung*: this is close to the English word 'tension'. Duit concludes:

> On the one hand, the logical structure of language leads to a way of thinking which is not appropriate in electricity. This appears to be true especially for the inability of most students to think of the electrical circuit in terms of a system. On the other hand the everyday meanings of concept names may hamper an understanding of the physical meaning... it is most important to bear in mind that the language teachers use in physics instruction is responsible for some further problems. (p 212)

In a comprehensive book about language in the science classroom, Sutton (1992) touches upon the issues of electrical language at several points, quoting the historical origins of many of the words as sources of confusion.

There is much that could be said here about the origins and remedies for electrical misconceptions. The electron-flow model itself is, I believe, responsible for much of the problem (Stocklmayer and Treagust, 1996; Stocklmayer and Treagust, 1994; Stocklmayer, Zadnik and Treagust, 1993). Once again, however, for the science communicator it is important to *know* that these ideas exist in the minds of the audience.

4.2. Alternative Conceptions in Chemistry

In chemistry, as in physics, there is a multitude of alternative conceptions which often surprise practising scientists. Areas such as chemical reactions, equilibrium, the mole concept, and the particulate nature of matter have all been found to have problems. Once again, I have selected two to describe here.

4.2.1. The Structure of Matter: Atoms and Molecules

Misconceptions about the structure of matter include ideas that matter exists between atoms, and that atoms are alive (because they move) (Griffiths and Preston, 1992). These researchers found that understanding molecular structure was very difficult, with Grade 12 students holding a great many misconceptions which had their origins in the world outside the classroom. For example, many believed that water molecules were made up of water, air, chlorine, nitrogen or minerals (p.615). 'Water has to go through many pipes before it reaches your tap so therefore it adds on more things...' (p. 618). Water molecules contain 'hundreds of thousands of atoms' and students had little idea of their real size or weight. Gaseous molecules are believed to be smaller than solid ones and weigh less.

Ball and stick models may be responsible for students believing that atoms resemble solid spheres with things inside. Certainly most students imagine their size in terms of the smallest thing that they can see - some fraction of a piece of dust, for example.

Griffiths and Preston (1992) in common with Brook, Briggs and Driver (1984) and Lee, Eichinger, Anderson, Berkheimer and Blakeslee (1993) found that students believe that macroscopic qualities apply equally to small particles: that they can, for example, become hot and even melt.

What happens to particles in processes such as evaporation, for example, is confusing to students. Stavy (1990) found many students did not understand that matter is conserved in evaporation and Bar and Travis (1991) found that 'the ideas that children have about air may be different from those of grown-ups' (p.380). Benson, Wittrock and Baur (1993) investigated understandings of the behaviour of air, in a flask which was then partially evacuated. They found that, for a very large sample of students who were all studying science, most had very confused conceptions of the behaviour of the air in the flask. More than 30% of those at university level (who were all studying chemistry) showed serious misconceptions.

When it comes to the structure of the atom, the planetary model, once learned, is exceptionally persistent. This problem was highlighted by Sandomir, Stahl and Verdi (1993) who stated that students:

> ...constructed an image of the atom that was essentially identical to their image and description of our solar system, even ignoring or distorting information from both the Bohr and quantum models to make these models fit their solar system model. (p. 22)

There is now, in some quarters, the thought that teaching the planetary model is a bad idea.

4.2.2. Electrochemistry, Oxidation and Reduction

Senior students are introduced to electrochemistry in terms of oxidative-reductive processes and often become confused about their nature. Garnett and Treagust (1992) and Garnett and Hackling (1993) identified a number of ideas held by senior high school students which were the result of teaching rather than pre-conceived notions. Oxidation and reduction processes were seen as independent and the assigning of oxidation states to various species was misunderstood. A striking area of confusion was the nature of current in electrolytes. The movement of ions was a source of puzzlement, current in general being regarded as a phenomenon involving free electrons flowing along the wire, then travelling through the salt bridge and the electrolytes. Some thought that electrons move through the solution by jumping on and off ions, and - well known to teachers - some had the idea that electricity in chemistry and physics is different. This 'compartmentalisation of science' was criticised by Garnett and Treagust (1992).

Many students in these studies ignored any water present as if it were inert. These findings were repeated in research with college students reported by Sanger and Greenbowe (1997), who also found confusion over the idea of a standard cell. Many probable causes of these ideas were identified by the researchers. The use of multiple definitions and models was a concern, as was inadequate prior knowledge. Both of these are relevant to science communication in general, as is the finding of Garnett and Treagust (1992, pp. 1095-1097) that language is one major source of confusion. Statements such as 'A salt bridge assists the flow of electrons' was interpreted by students to mean that 'positive ions in the salt bridge attracted electrons up and over to the other half cell'. Even a statement such as 'the salt bridge completes the circuit' was found to be interpreted as meaning that the bridge compensated for electrons removed during reduction.

'Ions carry the charge' is responsible for the image of the ions carrying electrons around. It is easy to see, with hindsight, how chemical language gives rise to strange ideas because, fundamentally, these learners have no real grasp of the nature of the processes. They manufacture pictures from the words of everyday - with disastrous results.

4.3. Alternative Conceptions in Biology

In biology, alternative conceptions have been identified in areas which include evolution, (many misconceptions here and a considerable volume of research) genetics, reproduction, homeostasis, circulatory systems, foodwebs, surface area to volume ratios, and principles of classification. I discuss two areas below that are related: they are respiration and photosynthesis.

4.3.1. Respiration and Photosynthesis

Photosynthesis is taught quite early in the average school syllabus and it is notable that it is usually presented in its entirety (including the chemical equation) right from the first time it is mentioned. The concept of photosynthesis as a process that uses the sun's energy to produce carbohydrates which later break down again to give energy for the plant's life processes requires quite sophisticated understanding to make sense of it. There is, in most texts, no sense of building the knowledge until a final picture can be made of the whole process. The degree of difficulty in describing the phenomenon alters little from junior to senior high school. Small surprise, therefore, that students find this process difficult to understand. Their difficulties have been documented consistently over the past two decades and various remedial approaches have been tested (see, for example, Haslam and Treagust, 1987; Semour and Longden, 1991. Anderson, Sheldon and Dubay (1990) reviewed the research in this area and found that similar misconceptions exist almost irrespective of the age of the students.

Eisen and Stavy (1988), for example, examined five central concepts related to photosynthesis, including oxygen release by plants and respiration. Their sample groups were high school and university students, some of whom were biology majors and some not. They found that 'adults, who did not study advanced courses in biology... show very little understanding of the essential role of photosynthesis in the ecosystem... It is clear that people do not appreciate the role of solar energy as the motive power of life.' (p. 211). Eisen and Stavy found confusion over the meaning of the word 'food' as related to plants, with over one third of the non-biology majors professing ignorance of what constituted a plant's food.

It seems to me that the term 'food' itself has real-world connotations that present problems immediately in the context of plants. Lumpe and Staver (1995) also found deep confusion in a group of high school biology students over what constituted 'food' for plants and where this food came from. Students identified water, minerals and so on as being food for plants, and failed to distinguish photosynthesis from nutrition.

The relationship between photosynthesis and respiration is even more problematic. Songer and Mintzes (1994, p. 630) quote beginning college majors in biology as saying 'Plants don't need oxygen. They can get oxygen from the water.' And 'Plants undergo cellular respiration but it is different from us. They want carbon dioxide'. A more muddled view is reflected by 'The plant needs carbon dioxide to synthesise energy. Light energy is the source of energy for the plant itself. Plants get food from water and nutrients in the soil'.

The literature about this research, in terms of identification of the alternative ideas and of efforts to remedy them, is enormous. Suffice to say here that it is important to be aware of these conceptions and to seek wherever possible to produce conceptual conflict - that is, to introduce information and observations that do not 'fit' the alternative conception. Equally important is to avoid language that unconsciously reinforces the previously held idea. An example of this occurred recently during research in an interactive science centre (Gilbert and Stocklmayer, 1998) during which many visitors explained that the 'Hand Battery' exhibit was demonstrating generation of electricity by the human body. Inspection of the graphic panel revealed that the text itself unconsciously was reinforcing this idea:

Making Batteries With Your Body

The simplest electric batteries are made from two different metals placed in salt water. Here, copper and aluminium are the two metals. The sweat on your hands provides the salt water. Each metal has a different chemical reaction with the salt water on your hands. This difference results in a flow of electricity from one metal plate to the other... When you connect the two metal plates with your hands, your body forms part of an unbroken circuit. Electricity flows through your body as well as through the different metals and the wires connecting them to the meter...

Congratulations! You have just made a simple hand battery. There is an electric current going through you...

Even visitors with experience of electricity transferred their knowledge of the body's internal electrical impulses to the phenomena observed at the exhibit:

I had heard the body generates electricity, but never tried it out before. It's interesting proof that we generate electricity. I was surprised it was a measurable quantity even though in microamps.' (Man in his 60s, 'electrically trained')

The problem with alternative conceptions is that they are remarkably tenacious and that people bend new information to suit existing beliefs. Solomon (1983) called them 'messy, contradictory and obstinately persistent'. Time after time, research has shown that remedial teaching has little effect. One of the main issues with such teaching has been shown to be the tendency for students to keep their classroom science separate from real world, 'common sense' knowledge (Fensham, 1993). Thus the teacher may illustrate quite graphically that the alternative conception cannot hold, but the information is stored in a region of the brain reserved just for science classes and not transferred to the belief system of the student.

The ability of students to retain several conflicting ideas each used in different circumstances has been noted as one of the disadvantages of our present system of education, particularly in science (Hewson and Hewson, 1991; Posner, Strike, Hewson, and Gertzog, 1982). The problem of alternative conceptions is just one facet of the problem of not remembering.

5. WHY THE PUBLIC IS 'TURNED OFF' SCIENCE AND CHOOSES NOT TO REMEMBER

In constructing a hypothetical audience, I chose to limit the argument to a Western group, albeit not a homogeneous one. The problems of other cultural views serve to complicate the picture further, introducing other ways of thinking which place more emphasis on intuitive and spiritual aspects, and the powers of the mind. When these dimensions are considered, the deficiencies of Western science become even more marked. Sandra Harding has discussed the political implications of this in two important books (Harding, 1993; Harding, 1998). Harding believes that we must question deeply the claims made for scientific objectivity, and for the independence of researchers, in the context of science in developing countries. (She questions also the notion that 'science' originated in Europe and that other sciences are 'merely technology'). There is not room here to pursue these implications but they reflect an extreme aspect of the problems described above. All people, Western or not, seek relevance for their understandings. A discipline which holds abstractness and objectivity as high virtues becomes, for the ordinary person, irrelevant. When that same discipline cloaks its doctrines in mystical, inaccessible language it becomes remote and elitist. Small wonder, then, that the excitement of science, its potential for positive change and its power to describe the natural world remain the province of a privileged few.

School students rarely see this privileged picture. Science is 'hard', 'boring', not important for real life. The emphasis is on remembering facts and formulae, not on seeing relevance all around them. For those who have little or no science education, which includes most adults, the barriers to understanding become too great to surmount.

How then can communicators begin to tackle this problem for diverse audiences with differing beliefs, interests and backgrounds?

6. THE PROBLEM FOR SCIENCE COMMUNICATION

The simple answer is to know the audience and to tailor the communication expressly for them. This is, of course not simple at all. It is almost impossible. To be effective, therefore, certain precautions have to be taken.

First, get rid of as much mathematics and formulae as you can. It's a terrible barrier, in the truly old-fashioned sense of striking terror into many of your audience.

Second, keep the language as straightforward as possible. Science Centres pitch their text at a *reading age* of 12. This is not to say that the form of words is at twelve-year old level–just the vocabulary. This is NOT 'talking down'. Even if you think the audience is more sophisticated, don't be tempted to make the language difficult. Science is hard enough without having to think about the words.

Third, think about the possibility of alternative conceptions. If you are serious about your communication, have a look at the research in the area. You will find it alarming!

Fourth, concentrate very hard on finding good introductory 'hooks'. If you don't place the communication in the world of the audience, they will not listen or read further.

Fifth, sixth, seventh: keep it simple, keep it simple, keep it simple.

These are very exacting demands. Science communication is a difficult profession - and I have, in this chapter, ignored the obvious need for scientists themselves to communicate more effectively with the public. But that is a whole story in itself!

National Centre for the Public Awareness of Science, Australian National University, Canberra ACT 0200, Australia

8. REFERENCES

Ancog, A.C. (1998). *Women in science*. Paper presented at the UNESCO Asia-Pacific Conference 'Science Issues for the 21ˢᵗ Century. Sydney, 2-6 December.

Anderson,C.W., Sheldon, T.S., & Dubay, J. (1990). The effects of instruction on college majors' conceptions of respiration and photosynthesis. *Journal of Research in Science Teaching, 27,* 761-776.

Bar, V., & Tavis, A.S. (1991). Children's views concerning phase changes. *Journal of Research in Science Teaching, 28,* 363-382.

Belenky, M.F., Clinchy, B., Goldberger, N.TR.,& Tarule, J.M. (1986) *Women's ways of knowing*. New York: Basic Books.

Benson, D.L., Wittrock, M.C., & Baur, M.E. (1993). Students' perceptions of the nature of gases. *Journal of Research in Science Teaching, 30,* 587-597..

Beynon, J. (1990). Some myths surrounding energy. *Physics Education, 25,* 314-316.

Bliss, J., & Ogborn, J. (1985). Children's choices of uses of energy. *European Journal of Science Education, 7,* 195-203.

Boyes, E., & Stanisstreeet, M. (1990). Misunderstandings of 'Law' and 'Conservation': a study of pupils' meanings for these terms. *School Science Review, 72* (258), 51-57.

Brook, A., Briggs, M., & Driver, R. (1984). *Aspects of secondary students' understanding of the particulate mature of matter*. Leeds: University of Leeds Children's Learning in Science Project. Centre for Studies in Science and Mathematics Education.

Butler Kahle, J. (1987). Images of science: The physicist and the cowboy. In B.J.Fraser and G.Giddings (Eds) *Gender issues in science education*. Perth: Curtin University of Technology (Research Seminar and Workshop Series), 1-12.

Campbell, L., & Krockover, G.H. (1992). *A qualitative study of preservice elementary teachers' developmental understanding of electricity and optics concepts.* Paper presented at the Annual Meeting of the National Association for Research in Science Teaching, Boston, MA.

Chambers, D.W. (1983). Stereotypic images of the scientist: The Draw-a-Scientist Test. *Science Education, 67,* 255-265.

Closset, J.L. (1983). Sequential reasoning in electricity. In *Proceedings of the First International Workshop, La Londe les Maures.* Paris: Editions du CNRS, 313-319.

Clough, E.E., & Driver, R. (1985). Secondary students' conceptions of the conduction of heat: Bringing together scientific and personal views. *Physics Education, 20,* 176-182.

Cobern, W.W. (Ed.) (1998). *Socio-cultural perspectives on science education.* Dordrecht: Kluwer.

Dierking, L.D., & Falk, J.H. (1994). Family behaviour and learning in informal settings: A review of the research. *Science Education, 78,* 57-72.

Driver, R., & Bell, B. (1986). Students' thinking and the learning of science: A constructivist view. *School Science Review, March,* 443-456.

Driver, R., Guesne, E., & Tiberghien, A. (Eds) (1985). *Children's ideas in science.* Milton Keynes: Open University Press.

Duit, R. (1985). Students' representations of the topological structure of the simple electrical circuit. In R. Duit, W. Jung, & C. von Rhöneck (Eds.), *Aspects of understanding electricity: Proceedings of an International Workshop.* Ludwigsburg, Germany: IPN - Kiel, 83-93.

Duit, R. (1987). Should energy be illustrated as something quasi-material? *International Journal of Science Education, 9,* 139-145.

Dupin, J., & Johsua, S. (1987). Conceptions of French pupils concerning electric circuits: Structure and evolution. *Journal of Research in Science Teaching, 24,* 791-806.

Eisen, Y., & Stavy, R. (1988). Students' understanding of photosynthesis. *The American Biology Teacher, 50,* 208 – 212.

Eylon, B., & Ganiel, G. (1990). Macro - micro relationships: the missing link between electrostatics and electrodynamics in students' reasoning. *International Journal of Science Education, 12,* 79-94.

Falk, J.H., Koran, J.J. Jr., & Dierking, L.D. (1986). The things of science: Assessing the learning potential of science museums. *Science Education, 70,* 503-508.

Fensham, P.J. (1993). Common sense knowledge: A challenge to research. *Australian Educational Researcher, 20,* (1). 1-20.

Finson, K.D., Beaver, J.B., & Cramond, B.L. (1995). Development and field test of a checklist for the Draw-a-Scientist Test. *School Science and Mathemnatics, 95.* 195-205.

Feynman, R.P. (1966). *The Feynman Lectures, Vol. 1.* New York: Addison Wesley.

Fredette, N., & Lochhead, J. (1980). Student conceptions of simple circuits. *The Physics Teacher, 18* (3), 194-198.

Gardner, H. (1993). *Multiple Intelligences: The theory in practice.* New York: Basic Books.

Garnett, P.J., & Hackling, M.W. (1993). Chemistry misconceptions at the secondary-tertiary interface. *Chemistry in Australia, March,* 117-119.

Garnett, P.J., & Treagust, D.F. (1992). Conceptual difficulties experienced by senior high school students of electrochemistry: Electrochemical (Galvanic) and electrolytic cells. *Journal of Research in Science Teaching, 29,* 1079-1100.

Gilbert, J.K., & Stocklmayer, S.M. (1998). Mental modeling in science and technology centres: What are visitors really doing? In S.Stocklmayer & T.Hardy (Eds.) *Proceedings of the international conference on learning science in informal contexts.* Canberra: Questacon, pp 16-32.

Griffiths, A.K., & Preston, K.R. (1992). Grade 12 students' misconceptions relating to fundamental characteristics of atoms and molecules. *Journal of Research in Science Teaching, 29,* 611-628.

Harding, S. (1991). *Whose science? Whose knowledge?* Milton Keynes: Open University Press.

Harding, S. (Ed.) (1993). *The 'racial' economy of science: Toward a democratic future.* Bloomington: Indiana University Press.

Harding, S. (1998). *Is science multicultural? Postcolonialisms, feminism and epistemologies.* Bloomington: Indiana University Press.

Haslam, F., & Treagust, D.F. (1987). Diagnosing secondary students' misconceptions of photosynthesis and respiration in plants using a two-tier multiple choice instrument. *Journal of Biological Education, 21,* 203-211.

Hermawati, S.A.W.(1998). *Gender in science: the case of Indonesia.* Paper presented at the UNESCO Asia-Pacific Conference 'Science Issues for the 21st Century. Sydney, 2-6 December.

Hewson, P. W., & Hewson, M.G.A. (1991). The status of students' conceptions. In R. Duit, F. Goldberg, & H. Niedderer (Eds.), *Research in physics learning: theoretical issues and empirical studies* Kiel: IPN.

Honey, P., & Mumford, A. (1986). *Using your learning skills.* Maidenhead, Berks.

Igelsrud, D. (1989). How living things obtain energy: A simpler explanation. *American Biology Teacher, 51,* 89-93.

Kirkwood, V., & Carr, M. (1989). A valuable teaching approach: Some insights from LISP (Energy). *Physics Education, 24,* 332-334.

Koran, J.J., Koran, M.L., Camp, B.D., & Donnelly, A.E. (1996). A summary of recent research and evaluation studies in the University of Florida program on learning in informal settings. *Visitor Behaviour, 11*(3), 5-8.

Lannes, D., Flavoni, L., & De Meis, L. (1998). The concept of science among children of different ages and cultures. *Biochemical Education, 26,* 199-204.

Lee, O., Eichinger, D.C., Anderson, C.W., Berkheimer, G.D., & Blakeslee, T.D. (1993). Changing middle school students' conceptions of matter and molecules. *Journal of Research in Science Teaching, 30,* 249-270.

Lumpe, A.T., & Staver, J.R. (1995). Peer collaboration and concept development: Learning about photosynthesis. *Journal of Research in Science Teaching, 32,* 71-98.

Jarvis, T. (1996). Examining and extending young children's views of science and scientists. In L.H. Parker, L.J. Rennie & B.J. Fraser (Eds). *Gender, science and mathematics: Shortening the shadow.* Dordrecht: Kluwer, 29-40.

Kelly, A. (1985). The construction of masculine science. *British Journal of Sociology of Education, 6,* 133-154.

Maloney, D. P. (1986). Rule governed physics - current in a series circuit. *Physics Education, 21,* 360-365.

McAdam, J.E. (1990). The persistent stereotype: children's images of scientists. *Physics Education, 25,* 102-105.

McDermott, L. (1993). How we teach and how students learn. *Australian & New Zealand Physicist, 30* (7), 151-163.

Millar, R., & King, T. (1993). Students' understanding of voltage in simple series electric circuits. *International Journal of Science Education, 15,* 339-349.

Moir, A., & Moir, B. (1998). *Why men don't iron: The real science of gender studies.* Harper Collins.

Ogborn, J. (1986). Energy and fuel: the meaning of 'the go of things'. *School Science Review, Sepember,* 30-35.

Osborne, R. J. and Gilbert, J.K. (1980). A method for investigating concept understanding in science. *European Journal of Science Education, 2,* 311-321.

Parker, L.H. (1996). *Transforming western science : Lessons from feminist scholarship.* Paper presented at a symposium at the Annual Meeting of the American Educational Research Association, New York, April.

Posner, G.J., Strike, K.A., Hewson, P.W., & Gertzog, W.A. (1982). Accommodation of a scientific conception: Toward a theory of conceptual change. *Science Education, 66,* 211-227.

Ramey-Gassert, L., Walberg, H.J.I., & Walberg, H.J. (1994). Reexamining connections: museums as science learning environments. *Science Education, 78*(4), 345-363.

Rennie, L. J., & McClafferty, T. P. (1996). Science centres and science learning. *Studies in Science Education, 27,* 53-98.

Rogan, J.M. (1988). Development of a conceptual framework of heat. *Science Education, 72,* 103-113

Ross, K.A. (1988). Matter scatter and energy anarchy. *School Science Review, March,* 438-445.

Sandomir, M.R., Stahl, R.J., & Verdi, M.P. (1993, April). *The atom is/is not a 'solar system' or an 'electron cloud': Metaphors as aids to and interferers of acquiring appropriate science content and conceptions - an information-constructivist perspective and preliminary findings.* Paper presented at the annual meeting of the National Association for Research in Science Teaching, Atlanta, GA.

Sanger, M.J., & Greenbowe, T.J. (1997). Common student misconceptions in electrochemistry: Galvanic, electrolytic and concentration cells. *Journal of Research in Science Teaching, 34,* 377-398.

Saxena, A. B. (1992). An attempt to remove misconceptions related to electricity. *International Journal of Science Education, 14,* 157-162.

Schibeci, R.A.(1986). Images of science and scientists and science education. *Science Education, 70,*

139-149.

Seymour, J., & Longden, B. (1991). Respiration – That's breathing isn't it? *Journal of Biological Education, 25*, 177-183.

Shipstone, D. M. (1984). A study of children's understanding of electricity in simple DC circuits. *European Journal of Science Education, 6*, 185-198.

Simpson, W.D., & Marek, E.A. (1988). Understandings and misconceptions of biology concepts held by students attending small high schools and students attending large schools. *Journal of Research in Science Teaching, 25*, 361-374.

Small, B. (1985). *Girl-friendly science: Avoiding sex bias in the curriculum*. London: Longman.

Solomon, J. (1983). Messy, contradictory and obstinately persistent: A study of children's out-of-school ideas about energy. *School Science Review, December*, 225-229.

Solomon, J. (1982). How children learn about energy, or Does the first law come first? *School Science Review, March*, 415-422.

Solomon, J., Black, P., Oldham, V., & Stuart, H. (1985). The pupils' view of electricity. *European Journal of Science Education, 7*, 281-294.

Songer, C.J., & Mintzes, J.J. (1994). Understanding cellular respiration: An analysis of conceptual change in college biology. *Journal of Research in Science Teaching, 31*, 621-637.

Stavy, R. (1990). Children's conceptions of changes in the state of matter: From liquid (or solid) to gas. *Journal of Research in Science Teaching, 27*, 247-266.

Stocklmayer, S.M., & Treagust, D.F. (1994). A historical analysis of electric currents in textbooks: A century of influence on physics education. *Science and Education, 3*, 131-154.

Stocklmayer, S.M., & Treagust, D.F. (1996). Images of electricity: How do novices and experts model electric current? *International Journal of Science Education, 18*, 163-178.

Stocklmayer, S.M., Zadnik, M.G., & Treagust, D.F. (1993). Teaching electricity: Is there a gender problem? In R.Schibeci (Ed), *Proceedings of the 18th Annual Conference of the Western Australian Science Education Association*, Perth: Murdoch University.

Sutton, C. (1992). *Words, science and learning*. Buckingham: Open University Press.

Tobin, K. (1990). Social constructivist perspectives on the reform of science education. *The Australian Science Teachers Journal, 36* (4), 29-35.

Trumper, R., & Gorsky, P. (1993). Learning about energy: The influence of alternative frameworks, cognitive levels and closed-mindedness. *Journal of Research in Science Teaching, 30*, 637-648.

Turney, J. (1998). *Frankenstein's footsteps*. New Haven: Yale University Press.

Viglietta, L. (1990). A more 'efficient' approach to energy teaching. *International Journal of Science Education, 12*, 491-500.

Warren, J.W. (1986). At what stage should energy be taught? *Physics Education, 21*, 153 – 156.

Wellington (1990) Formal and informal learning in science. *Physical education, 25*, 247-252.

Yager, R.E. (1991). The constructivist learning model: Towards reform in science education. *The Science Teacher, 58* (6), 52-57.

2. SCIENCE COMMUNICATION WITH THE PUBLIC: A CROSS-CULTURAL EVENT

1. INTRODUCTION

Scientist denies cancer cure quote May 8, 1998.

NEW YORK (AP) Nobel laureate James Watson denies telling a reporter a researcher whose experiments have rid mice of malignant tumors 'is going to cure cancer in two years.'

Watson, co-discoverer of the structure of DNA, was quoted as having made that prediction in a front-page story in Sunday's New York Times...

Watson, in a letter to the editor published in Thursday's Times, said he told Times science writer Gina Kolata at a dinner party six weeks ago that the drugs, endostatin and angiostatin, 'should be in the National Cancer Institute trials by the end of this year and that we would know, about one year after that, whether they were effective.'

Times spokesperson, Lisa Carparelli said, 'We're confident of the story we ran and don't wish to be in a position of quarrelling with a respected source and authority. We're glad we were able to let Dr. Watson further explain his view.'

This miscommunication between Watson and Kolata probably reflects differences between the community of scientists and the community of journalists. Key differences between the two cultures may have been veiled by the fact that both people spoke English, a language in which terms or phrases have multiple meanings and shift their meanings from context to context. Thus, an expression uttered in the context of scientists talking among themselves may have quite a different meaning on the front page of the New York Times. Perhaps both Watson and Kolata overlooked the cultural differences that defined their two communities.

This chapter focuses on the communication between different cultures, particularly between the culture of science and the culture of a public immersed in their everyday lives. Cultural anthropology suggests that science communication with the public is a cross-cultural event. If people do not clearly identify the cultures involved in the act of communicating, people risk the quagmire of miscommunication. A critical analytic understanding of the culture of Western science, and of the cultures of various audiences, is a prerequisite to effective science communication with the public. In the first part of this chapter, I summarise this prerequisite to effective communication, while in the second part, I describe effective communication in terms of culture brokering, illustrated in part by a case study of a recent Canadian science centre exhibit.

S.M. Stocklmayer et al. (eds.), Science Communication in Theory and Practice, 23–45.

2. A CULTURAL PERSPECTIVE ON WESTERN SCIENCE

Before we can think about the cultural aspects of science communication with the public, we first need to clarify what cultures and subcultures are. Then we need to understand how people cross cultural borders to communicate with each other. Last, we need to become conversant with anthropological research into the ease with which people cross cultural borders. In this section, I develop several key anthropological concepts that are applicable to the realm of science communication with the public.

2.1 Culture

Cultural anthropologists such as Geertz (1973, p. 5) have defined culture as

> an ordered system of meanings and symbols, in terms of which social interaction takes place.

This statement accurately describes the scientific community engaged in research, as scientists develop more accurate and sophisticated systems of meanings (theories, models, laws and principles, often expressed symbolically), and as they publish their manuscripts in journals (formal social interaction) to establish the validity of their ordered system of meanings. In addition to communicating through formal publications, social interactions take place in person, by e-mail, by telephone, at conferences, in the lab, in the field, and in bars or at other informal gatherings. According to Geertz's definition, science can be thought of as a culture with its own language and conventional ways of communicating for the purpose of social interaction within the community of scientists.

In an anthropological study of a high-energy physics community, Traweek (1992) described culture in a more detailed way:

> A community is a group of people with a shared past, with ways of recognizing and displaying their differences from other groups, and expectations for a shared future. Their culture is the ways, the strategies they recognize and use and invent for making sense, from common sense to disputes, from teaching to learning, it is also their ways of making things and making use of them (pp. 437-438, italics in the original).

By treating physicists as working within cultural borders, Traweek discovered some fascinating behaviour and bizarre communication by Japanese high-energy physicists as they negotiated between the subculture of their Japanese national physics community and the subculture of the international physics community. Traweek found that risk taking, power, culture, and subjectivity were all intermingled in ways that encouraged Japanese physicists to conform with their Japanese national physics community. This made it difficult for these Japanese physicists to cross the cultural border into the international community of high-energy physics. Japanese physicists were the target of pejorative humour, sarcasm, and cultural reprisals from their Japanese colleagues. Therefore, Japanese high-energy physicists had to cross into the culture of international physics with great care and subtlety by using humour, selected conformity and politics, so as not to offend their Japanese colleagues in high-energy physics. By recognising the cultural differences between Japanese high-energy physicists and international high-energy physicists, Traweek could better understand the otherwise bizarre communication among some Japanese physicists. Perhaps there is a lesson here for James Watson and Gina Kolata - they should have recognised science as a culture, a culture with borders that must be crossed if outsiders are to understand the communication conventions of that culture, and if insiders are going to communicate effectively with the public.

Consistent with both Geertz's and Traweek's definitions of culture, Phelan, Davidson and Cao (1991) suggested that culture be conceptualised as the

norms, values, beliefs, expectations, and conventional actions of a group. (p. 228)

This cogent definition helps to clarify how science is a cultural phenomenon. Science content can be subsumed under 'beliefs'. The communication conventions of scientists are guided by the norms, values, and expectations of the culture of science, and by the specific norms, values, and expectations of the specialty field of the scientist, that is, the his or her paradigm or scientific subculture. Other definitions of culture have guided research in science communication (for example, Banks, 1988; Bullivant, 1981; Ingle and Turner, 1981; Jordan, 1985; Maddock, 1981; Samovar, Porter and Jain, 1981; and Tharp, 1989). From these works one can establish the following list of attributes of culture: communication (psycho- and socio-linguistic), social structures (authority, participant interactions), customs, attitudes, values, beliefs, worldview, skills (psychomotor and cognitive), behaviour, and technologies (artefacts and know-how). In various studies, different attributes of culture have been selected as a focus on a particular interest in multicultural communication. The definition of Phelan *et al.* (1991) of culture (above) is advantageous because it has relatively few categories and they can be interpreted broadly to encompass all anthropological aspects of culture and subculture.

Just as there are paradigms (subcultures) within the culture of science, there are subgroups in everyday life, most commonly identified by race, language, and ethnicity, but which can also be defined by gender, social class, occupation and religion. Consequently, an individual simultaneously belongs to several subgroups; for instance, an oriental female Muslim physicist or a male middle-class Euro-American journalist. Large numbers and many combinations of subgroups exist due to the associations that naturally form among people in society. Each identifiable subgroup is comprised of people who generally embrace a defining set of norms, values, beliefs, expectations, and conventional actions. In short, each subgroup shares a culture, often called a 'subculture' to convey an identity with a subgroup. One can talk about, for example, the subculture of females, the subculture of the middle class, the subculture of the television media, or the subculture of a particular science museum.

2.2. Border Crossing

An everyday scenario will illustrate the difficulties people can encounter whenever they move between cultures or between subcultures:

> George and Gracie Smith flew from North America to Spain, physically crossing political borders, but not crossing cultural borders. After waiting 45 minutes in a restaurant for their dinner bill to arrive, George finally became vocally irate over the waiter's lack of service. The waiter, in turn, became hurtfully perplexed over the fact that his impeccable manners were not appreciated.

Misunderstandings can arise whenever one of the players does not recognise a cultural border that needs to be crossed for effective communication (Aikenhead, 1996).

People often cross cultural borders so easily that they do not realise they are even there - for example, when people move between the subculture of their friends and the subculture of their family home. But for people whose peer culture is vastly different from their home culture, transitions between friends and home can be psychologically hazardous and these transitions need to be negotiated carefully. Similarly problematic are the border crossings between humanist and scientific subcultures of Western society. This problem was identified by C.P. Snow (1964) in his classic *The Two Cultures*, pointing out the inability of people to speak to one another between these two cultures.

For people who feel at ease in both a humanist and scientific culture, however, border crossing is no problem. Border crossing for them is smooth. When people feel at ease like this, cultural borders seem invisible or nonexistent. It is when people begin to feel a degree of psychological discomfort with another subculture that border crossing becomes less smooth, and needs to be managed. Contributing to their discomfort may be some sense of disquiet with cultural differences or their unwillingness to engage in risk-taking social behaviour (depending on the situation, of course). When the self-esteem of people is in jeopardy (for instance, when playing badminton with players much better than they are or when participating in an unusual social occasion such as wearing a Halloween costume), border crossing could easily be hazardous. People may react in various ways to protect their egos. Even worse, if psychological pain is involved, avoidance is the natural response and border crossing becomes impossible. These descriptors of the ease of border crossing - smooth, manageable, hazardous, and impossible - are categories that Phelan *et al.* (1991) derived from their anthropological study of high school students who had to cross cultural borders between their homes and their school. This category system was helpful to Costa (1995) in her study of students' feelings of ease in science classes. The category system will be helpful in this chapter for understanding the role of a science communicator.

Border crossing into the culture of science can be made smoother for the public if science communicators know the culture of the everyday world of the public, and can contrast that culture with a critical analysis of the culture of science (its norms, values, beliefs, expectations, and conventional actions). But even more, a science communicator must consciously move back and forth between the public's everyday world and the scientists' world - switching norms explicitly, switching values explicitly, switching conceptualisations explicitly, switching expectations explicitly, and switching language conventions explicitly. The role of a science communicator is described in more detail later in this chapter.

2.3. Values and Norms

One principal component of any culture is its values and norms. Values and norms guide scientists whenever they decide between, for example, competing theories or competing experimental methodologies (Chubin, 1981). Values and norms are learned by the apprentice scientist and they become important aspects to his or her paradigm (Hawkins and Pea, 1987; Kuhn, 1970). Longino (1990) refers to this set of discipline-centred values as constitutive values (for example, parsimony, accuracy, open-mindedness, objectivity, etc.) In contrast to constitutive values, she points to the social context outside of science in which scientists live daily. She refers to these cultural values as contextual values. Her research documented cases in which these contextual values (rather than constitutive values) influenced the decisions taken by scientists over what 'facts' to believe. She concluded that science-as-practised (as opposed to science-as-imagined) is not value-neutral. The value-neutrality of science has also been falsified by other studies (Casper, 1980; Graham, 1981; Snow, 1987; Ziman, 1984). Those who believe in the neutrality of science contend that science is free of contextual values, not constitutive values.

Therefore, science communicators must be aware of the values and norms that are potentially inherent in the language conventions of scientists (their discursive practices). For instance, one constitutive value, scientific objectivity, is often communicated to the public through science textbooks. Textbooks, however, camouflage more subtle contextual values, for example, the value 'technoscience fix' (Carlsen et al., 1994; Factor and Kooser, 1981) - the idea that solutions to societal problems (such as water contamination) only require more scientific knowledge and more innovative technologies.

Moreover, when one examines the constitutive values within science, one discovers differences between the constitutive values espoused by scientists, and the constitutive values actually practised by scientists (Mitroff, 1974). For instance, scientists publicly revere objectivity but many rely on subjective hunches in the privacy of their labs. Holton (1978) explained this apparent conflict in values by distinguishing between two types of scientific activity - 'private science' and 'public science'. Each has a different social setting and therefore a different communication audience. Public science is communicated in journals, conference proceedings, textbooks and news releases, while private science is done in labs and communicated in personal notebooks, letters, e-mails, and informal conversations.

Private science communication is not necessarily guided by the same values and norms as communication in the public science arena. For example, subjectivity and closed-mindedness have advantages in private science but never in public science where objectivity and open-mindedness form the cultural expectations. It is interesting that research on scientists who analysed the Apollo moon rock samples in the early 1970's indicated that those who were held in high esteem by their colleagues used conflicting sets of values and norms (values and norms associated with public and private science), while those who were considered mediocre by their colleagues embraced only public science values and norms (Gauld, 1982).

Sociological research into present-day practices of scientists (e.g. Latour, 1987) concurs with Holton and Gauld. Scientific activity embraces two legitimate, dichotomous sets of values (norms and counter-norms). When public-science values and norms define the whole enterprise of science, they propagate myths about the nature of science because they hide the function of the private-science (sometimes guided by, for instance, subjectivity and closed-mindedness). Miscommunication is ripe whenever statements expressed in the social context of private science are repeated in the social context of public science. This distinction may shed light on the mis-communication between James Watson and Gina Kolata. Did Watson neglect to express his enthusiasm in the language conventions of public science (appropriate for the front page of the New York Times)? Did Kolata neglect to recognise Watson's expression as private science? Perhaps both failed to recognise the cultural border between the two subcultures - private science and public science - each with its own set of norms, values, beliefs, expectations, and conventional actions. When people do not see a cultural border to cross, they run the risk of miscommunicating.

Compared with scientists, the general public expresses an even wider array of values and norms, many of which conflict with those embraced by the culture of science. Cobern (1991) explored a way to identify clusters of values that seem to inform the public's general outlook on the world. Drawing upon the work of anthropologist Kearney (1984), Cobern investigated the way that people's worldviews may predispose them to being sympathetic or antagonistic toward the worldviews conveyed by much of Western science. Cobern and Aikenhead (1998) illustrated the ease of communication between a science teacher (Mr Hess) and students who generally shared his worldview toward nature (that is, orderly and understandable, governed by physical forces which can be fully understood by tearing nature apart and analysing the pieces - reductionism). On the other hand, students who possessed alternative worldviews toward nature, such as those formulated on aesthetic or spiritual orientations toward nature, had communication problems with Mr Hess. Worldview is a convenient concept that embodies fundamental

> presuppositions about what the world is really like and what constitutes valid and important knowledge about the world. (Cobern, 1996, p. 584)

Worldviews are basic culture-laden frameworks from which daily values and norms flow.

In summary, science communication with the public will be more effective when people recognise science as a culture having many subcultures (such as paradigms, as well as private and public subcultures). For both insiders and outsiders to the culture of science, cultural borders must be crossed before effective communication can take place between those two groups. These border crossings can be smooth, managed, hazardous, or impossible, depending upon the cultural differences experienced by individuals, and depending upon their resourcefulness and motivation to cross otherwise hazardous or impossible borders. Scientists are guided by a complex and dynamic set of cultural values and norms, as is the general public. Similarities and differences between these two groups may be better understood by considering their different worldviews.

3. THE CULTURE OF SCIENCE AND THE PUBLIC DOMAIN

The division between science and the general public manifests the theory/practice dichotomy endemic to Plato's eidos and praxis. This remnant of Greek culture continues to characterise Western thinking today. Western science tends to isolate itself in eidos (idealised pure knowledge), rendering itself superior to praxis (practical knowledge needed for action), according to Platonic Greek tradition. Therefore, understanding science communication predicated on that distinction becomes difficult for people for whom a theory-practice distinction does not exist. The problem can be eased somewhat by the communicator expressing the cultural features found on both sides of this cultural border.

How is scientific knowledge actually used outside the culture of science, in people's commonsense and professional life-worlds, that is, in praxis? A cherished myth in the culture of Western science is the belief that people can directly apply scientific knowledge to their everyday world (Aikenhead, 1980; Layton, 1991; Ryle, 1954; Solomon, 1983). Reality is much different. Based on case study research in the UK, Jenkins (1992) commented that using science in the everyday world is

> no more a straightforward application of the scientific knowledge acquired at school or in other formal contexts than technology is merely applied science. Rather it is about creating new knowledge or, where possible, restructuring, reworking and transforming existing scientific knowledge into forms which serve the purpose in hand. Whatever that purpose (political, social, personal, etc.), it is essentially concerned with action or capability, rather than with the acquisition of knowledge for its own sake. (p. 236)

This conclusion guides us in helping the general public negotiate what would otherwise be a hazardous border crossing between their everyday culture and the culture of science. One hazard is the fact that scientific knowledge must be deconstructed and then reconstructed in the context of everyday use (Layton, 1991). In the context of teaching science Layton, Jenkins, Macgill and Davey (1993) concluded:

> The nature of the transformation needed is not a matter which has hitherto commanded
> much attention from science teachers, although it has been a preoccupation of
> engineers for a century or more.....The essence of the problem is that the concepts
> developed by scientists in their quest for understanding [eidos] do not always map with
> exactitude onto the design parameters in terms of which practical action has to be
> planned [praxis]. As a result, for science to articulate with practice, some reworking is
> often required. (p. 129)

This communicative challenge has plagued science educators as they contemplate how to communicate with students over the use of science content outside of the classroom. The same challenge exists for all science communicators.

A case in point was a student (Melanie) who found border crossing into the culture of science hazardous when studying the topic of heat (Aikenhead, 1996). In spite of her high motivation to participate in hands-on group activities, Melanie could not cross the cultural border into the science of heat and temperature. Her difficulty may have arisen from her having a worldview at odds with the worldview generally embraced by science. Cobern (1996) argued,

> ...it is not that the students fail to comprehend what is being taught, it is simply that the
> concepts are either not credible or not significant' (p. 601) because 'for students it is
> aesthetic, religious, pragmatic, and emotional concepts that have scope and force with
> regard to nature. (p. 597)

Thus, a general distaste for mechanistic reductionist concepts (a central feature of a conventional scientific worldview) might explain why students such as Melanie choose not to integrate the scientific concepts of heat and temperature into their everyday notions of hot and cold (Kilbourn, 1980).

In the adult world of consumers, Layton et al. (1993) discovered that a scientific understanding of heat energy had no consequence to lay people managing domestic energy problems in their life-world. Layton and his colleagues seriously questioned the objective of science education to teach what is rarely usable in the everyday world. In the words of Wynne (1991):

> ordinary social life, which often takes contingency and uncertainty as normal and
> adaptation to uncontrolled factors as a routine necessity, is in fundamental tension with
> the basic culture of science which is premised on assumptions of manipulability and
> control. (p. 120)

These lessons from science education apply directly to the communication of science with the public. The more that someone's worldview differs from the one conveyed by Western science, the less smooth (the more impossible) will be their border crossing into the culture of science and, as a consequence, the more they challenge science communicators. Communicating effectively requires a knowledge of one's audience. Challenges can be met more realistically when we recognise that this communication is a cross-cultural event. Cultural gaps must be bridged, not just by content knowledge bridges (that is, the naive belief that the public only needs more accurate knowledge), but by bridges that communicate the norms, values, beliefs, expectations, and conventional actions of scientists (the culture of science).

Some useful distinctions among people in the public domain were identified by Ogawa (1998b) in the context of science education reform in Western countries. He described three types of orientations the public will assume towards science. His first type concerns whether a person understands science (science literacy versus science illiteracy). His second type of orientation addresses a more emotional aspect, whether a person supports science (a pro-science or anti-science position). Ogawa's third type of orientation deals with an ideological belief that scientific knowledge is the only valid form of knowledge to use in any context. This belief, often called 'scientism', privileges scientific knowledge over all other ways of knowing (Nadeau and Désautels, 1984). Thus, Ogawa's third type of orientation consists of pro-scientism versus anti-scientism. In short, Ogawa contends that people's stance toward science will be influenced by how they fit into these three types of orientations. Thus, their receptivity to, and engagement in, scientific communication will vary according to their literacy in science, their support of science, and their allegiance to scientism. Ogawa's scheme generated six orientations of people:

i.   science-literate, pro-science, pro-scientism folk
        ('science believers');
ii.   science-literate, pro-science, anti-scientism folk
        ('science contextualists');
iii.   science-literate, anti-science, anti-scientism folk
        ('authentic anti-scientists');
iv.   science illiterate, pro-science, pro-scientism folk
        ('science fanatics');
v.   science illiterate, pro-science, anti-scientism folk
        ('science vigilantes'); and
vi.   science illiterate, anti-science, pro-scientism folk
        ('neo anti-scientists').

These categories can sensitise science communicators to the challenges that face them and their Western audiences.

4. WESTERN SCIENCE AND NON-WESTERN CULTURES

Communication barriers are even more pronounced between Western science and non-Western cultures. Researchers have investigated the obstacles encountered when one teaches Western science to non-Western students. Their findings are highly relevant to science communication with the public. Because science tends to be a Western cultural icon of prestige, power, progress, and privilege, the culture of science tends to permeate the culture of those who engage it, with cultural assimilation being one possible negative consequence (Baker and Taylor, 1995; Dart, 1972; Jegede and Okebukola, 1991; MacIvor, 1995; Ogawa, 1995). This assimilation threatens indigenous cultures, thereby causing these people to experience Western science as a hegemonic icon of cultural imperialism (Battiste, 1986; Ermine, 1995; Linkson, 1998). Science communicators in the global village need to extend their cultural sensitivity to a public outside of Western culture.

The encroachment of Western culture occurs, in part, because it is hidden in the Trojan horse of Western science. Different cultures have reacted differently to this encroachment. Aboriginal, Japanese, and Islamic peoples represent three cultural groups that have fought against such assimilation. Each group is discussed here in turn. Emphasis is given to Aboriginal peoples because they are the most under-represented group in Western science. Nevertheless they must deal with Western scientists in the areas of health, land management, and ethics (MacIvor, 1995; Wolfe, Bechard, Cizek and Cole, 1992). Their perspective helps us understand the cultural borders that most people in the global village must cross before effective science communication can succeed.

4.1 Aboriginal Cultures

Knudtson and Suzuki (1992) documented various indigenous knowledge systems around the world that describe and explain nature. Aboriginals, they claimed, possess powerful knowledge systems that convey wisdom, a key element missing in Western science. Aboriginal knowledge about the natural world (Aboriginal science) contrasts with Western scientific knowledge in a number of other ways. The following summary of Aboriginal science is based on sensitive and scholarly analyses by Christie (1991), Ermine (1995), Kawagley (1990), Linkson (1998), McKinley (1996), Mitchie, Anlezark and Uibo, (1998), Peat (1994), Pierotti and Wildcat (1997), Pomeroy (1992), and Roberts and Wills (1998). They wrote about the Maori in Aotearoa (New Zealand), the original peoples of Australia, and the First Nations peoples on Turtle Island (America).

Aboriginal and Western science differ in their social goals: survival of a people versus the luxury of gaining knowledge for the sake of knowledge and for power over nature and other people. They differ in intellectual goals: to co-exist with mystery in nature by celebrating mystery versus to eradicate mystery by explaining it away. They differ in their association with human action: intimately and subjectively interrelated versus formally and objectively decontextualised. They differ in other ways as well: holistic Aboriginal perspectives with their gentle, accommodating, intuitive, and spiritual wisdom, versus reductionist Western science with its aggressive, manipulative, mechanistic, and analytical explanations.

The Western world has capitulated to a dogmatic fixation on power and control at the expense of authentic insights into the nature and origin of knowledge as truth (Ermine, 1995).

> They even differ in their basic concepts of time: circular for Aboriginals, rectilinear for scientists. (p. 102)

Aboriginal and scientific knowledge differ in epistemology. Pomeroy (1992) summarises the difference found on Turtle Island:

> Both seek knowledge, the Westerner as revealed by the power of reason applied to natural observations, the Native as revealed by the power of nature through observation of consistent and richly interweaving patterns and by attending to nature's voices. (p. 263)

Ermine (1995) contrasts the exploration of the inner world of all existence by his people with a scientist exploring only the outer world of physical existence. He concludes:

> Those who seek to understand the reality of existence and harmony with the environment by turning inward have a different, incorporeal knowledge paradigm that might be termed Aboriginal epistemology. (p. 103)

Along similar lines, Roberts and Wills (1998) compare a fundamental Maori ontological principle of 'whakapapa', an orientation to the past that connects a person to the creators of the land, with a Western scientific future orientation that embraces a preoccupation with matter and causal mechanisms.

Battiste (1986) explicates a Turtle Island epistemology by giving detail to what Pomeroy (1992) called 'naturequotes voices':

> A fundamental element in tribal epistemology [lies] in two traditional knowledge sources:
> 1. from the immediate world of personal and tribal experiences, that is, one's perceptions, thoughts, and memories which include one's shared experiences with others; and
> 2. from the spiritual world evidenced through dreams, visions, and signs which (are) often interpreted with the aid of medicine men or elders. (p. 24)

On the one hand, subculture science is guided by the fact that the physical universe is knowable through rational empirical means, albeit Western rationality and culture-laden observations (Ogawa, 1995); while on the other hand, Aboriginal knowledge of nature celebrates the fact that the physical universe is mysterious but can be survived if one uses rational empirical means, albeit Aboriginal rationality and culture-laden observations (Pomeroy, 1992). For example, when encountering the spectacular northern lights, Western scientists would ask, 'How do they work?' while the Waswanipi Cree ask, 'Who did this?' and 'Why?' (Knudtson and Suzuki, 1992). We can learn more about the culture of Western science the more we contrast it with other ways of knowing nature.

The norms, values, beliefs, expectations, and conventional actions of Aboriginal peoples contrast dramatically with the culture of Western science. Western science has been characterised as essentially mechanistic, materialistic, reductionist, empirical, rational, decontextualised, mathematically idealised, communal, ideological, masculine, elitist, competitive, exploitive, and impersonal (Fourez, 1988; Kelly, Carlsen and Cinningham, 1993; Rose, 1994; Snow, 1987). By comparison, Aboriginal sciences tend to be thematic, survival-oriented, holistic, empirical, rational, contextualised, specific, communal, ideological, spiritual, inclusive, cooperative, coexistent, and personal. Based on these two lists, Western science and Aboriginal sciences share some common features (empirical, rational, communal, and ideological). Consequently, it is not surprising that efforts are underway to combine the two knowledge systems into one field called 'traditional ecological knowledge' (Corsiglia and Snively, 1995). While a romanticised version of the peaceful coexistence of an Aboriginal with the environment should be avoided, Knudtson and Suzuki (1992) document the extent to which environmental responsibility is globally endemic to Aboriginal cultures. It is this quality that led Christie (1991), Pierotti and Wildcat (1997), Roberts, Norman, Minhinnick, Wihongi and Kirkwood, (1995), and Simonelli (1994) to define scientific ecology and sustainable Western science in terms of Aboriginal cultures. Simonelli (1994) quoted a Lakota ceremonialist's view of science and technology:

> This is not a scientific or technologic world. The world is first a world of spirituality. We must all come back to that spirituality. Then, after we have understood the role of spirituality in the world, maybe we can see what science and technology have to say. (p. 11)

Deloria (1992), also of the Lakota nation, challenged Western science's objectivity and validity when he spoke about improving the culture of science by getting scientists to adopt an Aboriginal sense of contextualised purpose.

Differences between the culture of Western science and the cultures of indigenous Aboriginal students help to explain the apparent reticence of students to learn about heat and temperature, or about any other Western science concept (Aikenhead, 1997; Schilk Arewa, Thomson and White, 1995; Sutherland, 1988). The cultural borders around Western science are seldom smooth for Aboriginal peoples. For instance, in an American study of third grade children, Schilk *et al.* (1995) concluded, that the perceptions Indian students had of scientists, largely dictated by popular media, were in direct conflict with their Iroquois values (p. 3):

> Interviewer: Do you think you could be a scientist?
> Client: That's not something Indians do. I couldn't hurt things or blow things up.

As long as Western prestige, power, progress, and privilege continue to affront the wisdom of traditional knowledge of the land, science communication worldwide will be challenged. More than ever before, science communicators in the 21st century will be engaged in helping both Western scientists and Aboriginal peoples communicate with each other.

4.2. Japanese Culture

Japanese people, for the most part, resisted the encroachment of Western culture fairly successfully until the mid 19th century, when they were forced by the threat of physical violence to open their country to American commerce and technology (Shelley, 1993). Western science followed in due course. Before this cultural invasion, Japanese people had a knowledge of nature, of 'shizen', which encompassed descriptive and explanatory elements as well as cosmological characteristics (Ogawa (1997):

Everything surrounding human life (for example, mountains, rivers, plants, trees, insects, fish or animals) has its own spirit, which can communicate with each other as well as with the people living there. Thus, the special feelings summarised by the 'one-bodiness', which means that human beings and every natural thing are one body in total, are felt by the Japanese. (p. 176)

Elsewhere Ogawa (1998a, p. 158) asserts that most Japanese feel and are familiar with such spirits. Fortified by a feeling of animism, Japanese people cannot regard natural things as mere objects of value-free inquiry, as Western scientists tend to do. Although the word 'observe' is translated into Japanese by 'kansatsu', that is not an accurate translation because kansatsu connotes a close spiritual-like relationship, much different from the objective relationship presupposed by the Cartesian dualism (the mind/matter dichotomy) that forms a cornerstone to a Western science worldview (Kawasaki, 1996b).

The observer-object relationship connoted by kansatsu is not unique to Japanese culture. Most Aboriginal languages lack a verb 'to observe' with a Cartesian connotation. For example, Canadian Plains Cree use the verb 'kanawapamew' to indicate a visual connection to an animate object. The verb changes as the human sense changes to hearing or touching, and as the classification of an object changes to inanimate - connoting a very different relationship. Similarly, feminist writers have critiqued Western science for its hegemonic Cartesian discourse (Rose, 1994; Scantlebury, 1998). For example, Barbara McClintock's highly successful scientific research, described in Keller's (1983) *A Feeling for the Organism*, supports the view that alternative observer-object relationships can successfully advance Western scientific knowledge.

Discourse is highly dependent upon one's worldview. Because discourse is central to science communication with the public, we need to be sensitive to our own language, to the language of Western science, and to the language of our audience. For instance, to describe scientists observing a distant galaxy requires a different discourse depending upon the audience. An effective communicator should be able to acknowledge the audience's conventional meaning of 'to observe' and then be able to articulate the cultural border that needs to be crossed in order to appreciate what scientists have done when they have 'observed' that galaxy. In short, the science communicator must help an audience cross the cultural border into science sufficiently to engage in the act of communicating.

Returning to the topic of Japanese culture, the accelerated acculturation of Western science that followed the American incursion of 1863 was consciously controlled by the Japanese intelligentsia. They were very much aware of the cultural border between Western nature and Japanese shizen (Kawasaki, 1996a). The cultural border is identified today by such expressions as, 'I may wear a Western suit, but I have a bamboo heart'.

The degree to which Japanese people embrace Western materialism and its ideology of progress, is the degree to which Western assimilation seems to have succeeded (Suzuki and Oiwa, 1996). However, Japanese people have exhibited a type of acculturation of Western science that protects their bamboo hearts. They transform this foreign element into something quite new.

> The paradox is that Japanese culture through its long history has been able to adopt various components of foreign culture without losing its own identity. (Ogawa, 1998a, pp. 142-143)

Thus, Western science is transformed into something different, even though it is still called Western science or 'neo-science' by Ogawa (1997).

This information helps to explain Traweek's (1992) observations of the difficulties experienced by high-energy Japanese physicists when they attempted to move between the subculture of Japanese physics into the subculture of international (Western) high-energy physics, described earlier in this chapter. Perhaps the Japanese high-energy physicists were negotiating the boundary between 'neo-science' and Western science by switching worldviews and values/norms as they crossed the border. Ogawa's analysis also points out that if we wish to communicate Western science to a Japanese public, we must cross two cultural borders: from Western science to a transformed Western science (neo-science), and then to the everyday culture of Japanese society. Corresponding challenges exist for science communicators in societies other than Japan.

Even within Western cultures, Western science was shown (earlier in this chapter) to be transformed into a different knowledge system whenever science is used for practical action (Jenkins, 1992; Layton, 1991; Layton et al., 1993). The similarity to the Japanese transformation of Western science into neo-science is striking.

In summary, science communication is much more complex than transmitting scientific information. One needs to respond to the multiple cultures or subcultures involved, not only within the culture of Western science, but within one's audience. If a science communicator does not realise the culture-laden nature of science as practised in Euro-American institutions (Western science), and the culture-laden nature of its discourse, then he or she runs very high risks of creating misunderstandings in an audience. If a science communicator does not critically analyse his or her own linguistic conventions, and those of his or her audience, then he or she runs very high risks of mis-communicating with the public. Sensitivity to our discourse is fundamental to science communication with the public.

4.3. Islamic Cultures

The encroachment of Western (Greek) thought into Islam has a very different history from either the attempted colonisation of Aboriginal peoples or the acculturation of Japanese people. The history of Islamic science during the 8th to 12th centuries is largely characterised by a multicultural synthesis involving knowledge and technique from China, India, Greece, and Arabian nations (Krugly-Smolska, 1992).

Today, Islamic nationalism has created several views toward Islamic science, with different sects (or movements) defining Islamic science differently. Sardar (1997) described five competing views. One major issue in this debate is an epistemological issue. It concerns re-establishing the relationship between revelation (knowledge found in the Qur'an) and reason (inductive empirical knowledge).

> Revelation in Islam is above reasoning, but not above reason. Neither is reason above revelation. This subtle relationship was destroyed when Greek thought became dominant in Muslim societies. (Sardar, 1989, p. 13)

One of Sardar's (1997) five categories of Islamic science - 'mystical fundamentalism' - was the object of interest in Irzik's (1998) analysis of Islamic science in modern Turkey. Irzik contended that in their search for an Islamic science, 'radical intellectuals' reject Western science in terms that parallel critiques by Japanese people (Kawasaki, 1996a) and by Aboriginals (Christie, 1991; Ermine, 1995).

> These [radical] intellectuals also criticize industrialization on the grounds that production for the sake of ever more profits has turned human beings into mere puppets of Capitalist consumer society manipulated by mass media, deprived them of their religious-spiritual values, and enslaved them to the greed for material wealth. (Irzik, 1998, p. 167)

A significant segment of the global public will likely perceive scientific knowledge generated in Western institutions as laden with Western values and morally bankrupt. Communicators of science must keep in mind the various socially constructed realities of different publics.

For instance, Sardar's (1989) balance of revelation and reason has been replaced in some Islamic quarters by a radical form of Islamic science in which a hierarchy of revelation over reason exists. This hierarchy is dedicated to the principle of unity in which Allah, humankind, and nature (bodies and souls included) exist 'in harmony with the natural order of things' (Irzik, 1998, p. 173).

No matter what the culture, a fundamentalist public anywhere will present a great challenge to effective science communication worldwide. Even in Western cultures, for instance, Ogawa (1998b) pointed out that some of the public (science believers, science fanatics, and neo anti-scientists - described earlier) embrace a pro-scientism ideology called 'scientific fundamentalism' by Sardar (1997). This fundamentalism will be of particular interest to those science communicators who catch themselves inadvertently communicating this fundamentalist scientism without being aware of the ideological baggage attached to their communication, and therefore being surprised by the negative reaction from their audience.

4.4. *Summary*

Communication of science with the public will occur in various cultural contexts each one populated by a different public. Not only are there diverse publics to be considered, but there are pluralistic sciences to be acknowledged. Western science has tended to dominate Aboriginal science, Japanese science, and Islamic science, not because of any intellectual or moral superiority but because Western science is embedded in a culture that has colonised large portions of the planet. Western science has been invested with much more authority then, for instance, everyday commonsense science, not because Western science is necessarily more valid in that context, but because its culture is associated with prestige, power, progress, and privilege. A question of truth is hybridised with a question of social or political privilege. These are some of the ideological features to science communication with the public of which a communicator must be aware.

5. CULTURE BROKERING

Challenges to effective science communication with the public have accumulated throughout this chapter. Potential solutions to these challenges were suggested. These solutions had common features in that they recognised the cultural nature of any science (Western or otherwise), the need to cross cultural borders when communicating science, and the need to be sensitive to the subcultures of the audience. They also recognised the need to acknowledge such cultural components as values, norms, ideologies, histories, epistemologies, and linguistic conventions, on both sides of the cultural border.

Stairs (1995) referred to people who facilitated Canadian First Nations peoples movement between Aboriginal and Euro-Canadian society as 'culture brokers.' A culture broker helps people move back and forth between cultures and helps them resolve any conflicts that might arise.

Communicating science to the public has traditionally been a process of transmitting scientific facts, principles, and triumphs (Dierking and Martin, 1997). This has largely been a one-way process (Layton et al, 1993). By re-conceptualising this communicating process as a two-way cross-cultural event, and by taking on the role of culture broker, a communicator's task changes fundamentally. A culture brokering science communicator acknowledges and respects the cultural perspective of his or her audience, a cultural perspective that has norms, values, beliefs, expectations, and conventional actions, some of which may conflict with those of Western science. The audience's indigenous science is neither ignored nor marginalised. A culture broker will identify the cultural border that separates the public's indigenous culture from the culture of Western science. In addition, the cultural nature of Western science is established, perhaps through an explication of some of its cultural features.

These aspects of culture brokering form a foundation for increasing the effectiveness of communicating Western science to the public; in other words, for increasing the ease with which the public can cross the cultural border into Western science, enough to participate in the communication intended by the science communicator.

This cross-cultural event is made even smoother when a science communicator consciously and explicitly moves back and forth between the culture of Western science and the cultures of the audience (audiences are often multicultural). This can be accomplished verbally, by labeling each side of the cultural border with some type of linguistic marker. This might be achieved, for example, by referring to a group of high-energy physicists as 'a tribe of physicists'; by relating stories that serve as defining moments of contrast, such as. the newspaper story that began this chapter; and most of all, by making it overtly clear which culture we are communicating in at any given moment, and by making it overtly clear when we cross into another culture. Visual and auditory cues constitute the creative substance of communication while humour is often its winning style.

A realistic goal for culture brokers is to make transitions across borders smoother for our audience by transforming: impossible borders into hazardous ones, hazardous borders into manageable ones, or manageable borders into smooth ones.

When crossing cultural borders, we invariably switch norms, values, beliefs, expectations, and conventional actions. This switching is done overtly for more effective communication. Both the communicator and the audience are aware of the critical changes in language conventions, in epistemology, in worldviews, and in ideology, that accompany the cultural border crossing event. This awareness defines the goal of a culture brokering science communicator.

6. A CASE STUDY OF SCIENCE COMMUNICATION

In 1997 at the Ontario Science Centre in Toronto, a radically different interactive exhibition opened. It is called *A Question of Truth.* It makes explicit the intimate relationship between scientific knowledge and social responsibility, in a cultural and political context (Pedretti, Mclaughlin, Macdonald and Kithinji, 1998).

The public is expected to learn more about their own views of science and to explore the culture and practice of science-in-action, not idealised science. *A Question of Truth* illustrates some of the key features identified in this chapter that contribute to effective communication with the public. The title of the exhibit expresses irony in that its content demonstrates a socially constructed truth, not an absolute truth inherent in scientism.

The exhibit engages people in three themes: frames of reference, biases in society and science, and science and the community. Pedretti and her colleagues (1998) described these themes this way:

> The exhibit ... is designed to examine several questions about the nature of science itself, how ideas are formed and how cultural and political conditions affect the actions of individual scientists. Practitioners of science are portrayed as having a point of view, one which is derived from personal, cultural and political aspects of their lives. The ultimate findings of science are shown to be human products of our society. The exhibit traces lines of bias throughout our western scientific history. Ideas which have

remained unquestioned for centuries are examined anew for their roots in human prejudice. (p. 4)

For instance, in the section 'Point of View' (a frame of reference theme), visitors are introduced to navigational beliefs of Pacific islanders, two views of the solar system, and different calendar systems. The exhibit constantly poses the question, 'Can you accept points of view different from yours?' Some science literate visitors became rather surprised when they learned that scientific (sun-centred) models of the solar system are not as useful as earth-centred models when it comes to navigating or creating calendars.

The education guide to *A Question of Truth* (OSC, 1997) warns educators who are planning a field trip to the new exhibit:

Please be aware that certain [sub]exhibits such as the Science & Prejudice video, Sex & Science, the Confinement Box, Slavery; Who is Civilized? and Speak Up! present some controversial views. Powerful questions and feelings could arise among both adults and young people. (p. 6)

Reactions have indeed been powerful. Pedretti *et al.* (1998) assessed the impact of the exhibit and found that

deconstructing long standing and deeply entrenched views of science can create tension and dissonance. (p 25)

A very small minority (8%) thought that *A Question of Truth* had no place in a science centre (perhaps people belonging to Sardar's 'scientific fundamentalism' or to Ogawa's 'pro-scientism' groups?). One main reason was the fact that *A Question of Truth* challenged their stereotype images of science by portraying

science as a human endeavour (and therefore socially and culturally bound, subject to error, differing views and values. (p. 19)

Pedretti and her colleagues raised the following point in the conclusion to their study of the public's interaction with *A Question of Truth.*

Science centre exhibits need to consider a more contextualized approach to their portrayal of science; one which recognizes the contributions of many cultures to our understanding of the natural world, and one which recognizes science as a human activity: value laden and contextually bound. This kind of exhibitry needs to be developed in addition to, or in tandem with, traditional phenomenon-based exhibits. (p. 25)

This conclusion addresses several features of culture brokering in science communication. Every culture develops a science that fits its unique needs (Feyerabend, 1988). Western science is but one way of rationally perceiving reality (Ogawa, 1995). Culture brokering requires a pluralistic understanding of science–a multi-science perspective.

Every science is embedded in a culture, including Western science. Any science is necessarily a social activity guided by norms, values, beliefs, expectations, and conventions with which social action takes place. These features can be contrasted with those held by a particular public with whom one is communicating a science.

Although *A Question of Truth* gave emphasis to a critical examination of Western science, the exhibit did attend to participants expressing their own views. Border crossing was facilitated by the questions posed and by the contexts in which visitors confronted those questions. However, the participants by and large were left to their own to handle any cultural conflict that may have arisen. Although interactive science centres tend to be more conducive to smoother border crossings than the more traditional museum displays, these centres, at present, still require the public to negotiate cultural borders by themselves. The public needs culture brokering communicators.

7. CONCLUSION

This chapter began with a miscommunication between scientist James Watson and journalist Gina Kolata. That miscommunication seems ironic because it was Watson himself who did so much towards communicating science effectively to the public when he published *The Double Helix* in 1968. His readers were made aware of the paradigms of practice, the cultural metaphors, the social conventions, and the competitive struggles that characterise science-in-action. His book lay bare for public scrutiny many of Western science's cultural features, even more vividly than subsequent scholarly treatises on the social construction of science (e.g. Latour, 1987; Latour and Woolgar, 1979; Longino, 1990). Watson portrayed fellow scientists as developing an ordered system of meaning and symbols (a helical model for DNA) in the context of social interactions. These social interactions were so poignantly portrayed that some people thought the book should be X-rated.

Latour (1987) criticises journalists who report every new development in technoscience as a breakthrough in the progress of humanity. He points out that they fail to communicate science culture effectively to the public by not, for instance, identifying the ideology of scientism associated with 'technoscience breakthroughs' by asking, for example, 'progress for whom?' A critical analytic understanding of the culture of Western science is a prerequisite to effective communication with the public. Western scientific knowledge and technique must be seen as socially constructed within paradigms of practice, and socially determined by cultural metaphors and conventions, by economic interests, and by competition for privilege and power.

This prerequisite knowledge about Western science was informed in this chapter by such anthropological concepts as culture, subculture, assimilation, acculturation, worldview, and ease of border crossing; and by related concepts such as constitutive and contextual values, norms and counter-norms, public and private science, ideology, and epistemology. These concepts were introduced throughout the first part of the chapter.

Because science is necessarily embedded in a culture, science does not transfer easily into other cultures, including the subculture of everyday praxis in Western nations. This problematic transferability was amplified by communication problems that arose when Western science was taught to a non-Western public. Western science, with its own set of norms, values, beliefs, expectations, and conventional actions, turns out to be only one way of making sense of nature. Not only do we science communicators need to be sensitive to multicultural audiences, but from time to time we will need to consider multiple sciences as well.

Sensitivity and knowledge are prerequisites to becoming an effective culture broker who can help audiences cross the cultural border into Western science, smoothly enough to engage with the science communicator. Culture brokering will be a new role for most communicators. It will take extended practice to cultivate and perfect.

College of Education, University of Saskatchewan, Saskatoon, Canada

8. REFERENCES

Aikenhead, G.S. (1980). *Science in social issues: Implications for teaching*. Ottawa, Ontario: Science Council of Canada.

Aikenhead, G.S. (1996). Science education: Border crossing into the subculture of science. *Studies in Science Education, 27*, 1-52.

Aikenhead, G.S. (1997). Toward a First Nations cross-cultural science and technology curriculum. *Science Education, 81*, 217-238.

Baker, D., & Taylor, P.C.S. (1995). The effect of culture on the learning of science in non-western countries: The results of an integrated research review. *International Journal of Science Education, 17*, 695-704.

Banks, J.A. (1988). *Multiethnic education*. 2nd ed. Boston, MA: Allyn & Bacon.

Battiste, M. (1986). Micmac literacy and cognitive assimilation. In J. Barman, Y. Herbert, & D. McCaskell (Eds.), *Indian Education in Canada, Vol. 1: The Legacy*. Vancouver, BC: University of British Columbia Press, 23-44.

Bullivant, B.M. (1981). *Race, Ethnicity and Curriculum*. Melbourne, Australia: Macmillan.

Carlsen, W., Kelly, G., & Cunningham, C. (1994). Teaching ChemCom: Can we use the text without being used by the text? In J. Solomon & Glen Aikenhead (Eds.), *STS education: International perspectives on reform*. New York: Teachers College Press, pp. 84-96.

Casper, B.M. (1980). Public policy decision making and science literacy. In D. Wolfle et al. (Eds.), *Public policy decision making and scientific inquiry: Information needs for science and technology (Report No. NSF-80-21-A6)*. Washington, DC: National Science Foundation.

Christie, M.J. (1991). Aboriginal science for the ecologically sustainable future. *Australian Science Teachers Journal, 37* (1), 26-31.

Chubin, D.E. (1981). Values, controversy, and the sociology of science. *Bulletin of Science, Technology, & Society, 1*, 427-436.

Cobern, W.W. (1991). *World view theory and science education research, NARST Monograph No. 3*. Manhattan, KS: National Association for Research in Science Teaching.

Cobern, W.W. (1996). Worldview theory and conceptual change in science education. *Science Education, 80*, 579-610.

Cobern, W.W., & Aikenhead, G.S. (1998). Cultural aspects of learning science. In B.J. Fraser & K.G. Tobin (Eds.), *International handbook of science education*. Dordrecht, The Netherlands: Kluwer Academic Publishers, 39-52.

Corsiglia, J., & Snively, G. (1995). Global lessons from the traditional science of long-resident peoples. In G. Snively & A. MacKinnon (Eds.), *Thinking globally about mathematics and science education*. Vancouver, Canada: University of British Columbia, Centre for the Study of Curriculum and Instruction, 25-50.

Costa, V.B. (1995). When science is 'another world': Relationships between worlds of family, friends, school, and science. *Science Education, 79*, 313-333.

Dart, F.E. (1972). Science and the worldview. *Physics Today, 25* (6), 48-54.

Deloria, V. (1992). Relativity, relatedness and reality. *Winds of Change, (Autumn)*, 35-40.

Dierking, L.D., & Martin, L.M.W. (Guest Eds.) (1997). Special issue: Informal science education. *Science Education, 81* (6).

Ermine, W.J. (1995). Aboriginal epistemology. In M. Battiste & J. *Barman (Eds.), First* Nations education in Canada: The circle unfolds. Vancouver, Canada: University of British Columbia Press, 101-112.

Factor, L., & Kooser, R. (1981). *Value presuppositions in science textbooks.* Galesburg, IL: Knox College.

Feyerabend, P. (1988). *Against method.* London: Verso Publishers.

Fourez, G. (1988). Ideologies and science teaching. *Bulletin of Science, Technology, and Society, 8*, 269-277.

Gauld, C. (1982). The scientific attitude and science education: A critical reappraisal. *Science Education, 66*, 109-121.

Geertz, C. (1973). *The interpretation of culture.* New York: Basic Books.

Graham, L.R. (1981). *Between science and values.* New York: Columbia University Press.

Hawkins, J., & Pea, R.D. (1987). Tools for bridging the cultures of everyday and scientific thinking. *Journal of Research in Science Teaching, 24*, 291-307.

Holton, G. (1978). The scientific imagination: Case studies. Cambridge: Cambridge University Press.

Ingle, R.B., & Turner, A.D. (1981). Science curricula as cultural misfits. *European Journal of Science Education, 3*, 357-371.

Irzik, G. (1998). Philosophy of science and radical intellectual Islam in Turkey. In W.W. Cobern (Ed.), *Socio-cultural perspectives on science education: An international dialogue.* Dordrecht, Netherlands: Kluwer Academic Publishers, 163-179.

Jegede, O.J., & Okebukola, P.A. (1991). The effect of instruction on socio-cultural beliefs hindering the learning of science. *Journal of Research in Science Teaching, 28*, 275-285.

Jenkins, E.W. (1992). School science education: Towards a reconstruction. *Journal of Curriculum Studies, 24*, 229-246.

Jordan, C. (1985). Translating culture: From ethnographic information to education program. *Anthropology and Education Quarterly, 16*, 104-123.

Kawagley, O. (1990). Yup'ik ways of knowing. *Canadian Journal of Native Education, 17* (2), 5-17.

Kawasaki, K. (1996a). The concepts of science in Japanese and western education. *Science & Education, 5*, 1-20.

Kawasaki, K. (1996b, September). *Kansatsu: The way to produce the Japanese eureka situation.* A paper presented at the international symposium on Culture Studies in Science Education, Ibaraki University, Mito, Japan.

Kearney, M. (1984). *World view.* Novato, CA: Chandler & Sharp Publishers, Inc.

Keller, E.F. (1983). *A feeling for the organism.* San Francisco: W.H. Freeman and Co.

Kelly, G.J., Carlsen, W.S., & Cunningham, C.M. (1993). Science education in sociocultural context: Perspectives from the sociology of science. *Science Education, 77*, 207-220.

Kilbourn, B. (1980). World views and science teaching. In H. Munby, G. Orpwood, & T. Russell (Eds.), *Seeing curriculum in a new light: Essays from science education.* Toronto: OISE Press, 34-43.

Knudtson, P., & Suzuki, D. (1992). *Wisdom of the elders.* Toronto, Canada: Stoddart.

Krugly-Smolska, E. (1992). A cross-cultural comparison of conceptions of science. In G.L.C. Hills (Ed.), *History and philosophy of science in science education, Vol. I.* Kingston, Ontario, Canada: Faculty of Education, Queen's University, 583-593.

Kuhn, T. (1970). *The structure of scientific revolutions (2nd Ed.).* Chicago: University of Chicago Press.

Latour, B. (1987). *Science in action.* Cambridge, MA: Harvard University Press.

Latour, B., & Woolgar, S. (1979) . Laboratory life: The social construction of scientific facts. London: Sage.

Layton, D. (1991). Science education and praxis: The relationship of school science to practical action. *Studies in Science Education, 19*, 43-79.

Layton, D., Jenkins, E., Macgill, S., & Davey, A. (1993). *Inarticulate science? Perspectives on the public understanding of science and some implications for science education.* Driffield, East Yorkshire, UK: Studies in Education.

Linkson, M. (1998, July). *Cultural and political issues in writing a unit of Western science appropriate for primary aged Indigenous students living in remote areas of the Northern Territory.* A paper presented to the 47th annual meeting of the Australian Science Teachers Association, Darwin, Australia.

Longino, H.E. (1990). *Science as social knowledge: Values and objectivity in scientific inquiry.* Princeton, NJ: Princeton University Press.

MacIvor, M. (1995). Redefining science education for Aboriginal students. In M. Battiste & J. Barman (Eds.), *First Nations education in Canada: The circle unfolds.* Vancouver, Canada: University of British Columbia Press, 73-98.

Maddock, M.N. (1981). Science education: An anthropological viewpoint. *Studies in Science Education, 8,* 1-26.

McKinley, E. (1996). Towards an Indigenous science curriculum. *Research in Science Education, 26,* 155-167.

Michie, M., Anlezark, J., & Uibo, D. (1998, July). *Beyond bush tucker: Implementing Indigenous perspectives through the science curriculum.* A paper presented to the 47th annual meeting of the Australian Science Teachers Association, Darwin, Australia.

Mitroff, I.I. (1974). Norms and counter-norms in a selected group of the Apollo moon scientists: A case study of the ambivalence of scientists. *American Sociology Review, 39,* 579-595.

Nadeau, R., & Désautels, J. (1984). *Epistemology and the teaching of science.* Ottawa, Canada: Science Council of Canada.

Ogawa, M. (1995). Science education in a multi-science perspective. *Science Education, 79,* 583-593.

Ogawa, M. (1997). The Japanese view of science in their elementary science education program. In K. Calhoun, R. Panwar, & S. Shrum (Eds.), *International Organization for Science and Technology Education 8th symposium proceedings, Vol. 2: Policy.* Edmonton, Canada: University of Alberta, 175-179.

Ogawa, M. (1998a). A cultural history of science education in Japan: An epic description. In W.W. Cobern (Ed.), *Socio-cultural perspectives on science education: An international dialogue.* Dordrecht, Netherlands: Kluwer Academic Publishers, 139-161.

Ogawa, M. (1998b). Under the noble flag of 'developing scientific and technological literacy.' *Studies in Science Education, 31,* 102-111.

Ontario Science Centre. (1997). *A question of truth: Education guide.* Toronto, Canada: Ontario Science Centre.

Peat, D. (1994). *Lighting the seventh fire.* New York: Carol Publishing Group.

Pedretti, E., McLaughlin, H., Macdonald, R., & Kithinji, W. (1998, May). *A question of truth: Exploring the culture and practice of science through science centres.* A paper presented at the annual meeting the Canadian Society for Studies in Education, Ottawa, Canada.

Phelan, P., Davidson, A., & Cao, H. (1991). Students' multiple worlds: Negotiating the boundaries of family, peer, and school cultures. *Anthropology and Education Quarterly, 22,* 224-250.

Pierotti, R., & Wildcat, D.R. (1997). The science of ecology and Native American tradition. *Winds of Change, (Autumn),* 94-97.

Pomeroy, D. (1992). Science across cultures: Building bridges between traditional Western and Alaskan Native sciences. In G.L.C. Hills (Ed.), *History and philosophy of science in science education, Vol. II.* Kingston, Ontario, Canada: Faculty of Education, Queen's University, 257-268.

Roberts, R.M., Norman, W., Minhinnick, N., Wihongi, D., & Kirkwood, C. (1995). Kaitiakitanga: Maori perspectives on conservation. *Pacific Conservation Biology, 2,* 7-20.

Roberts, R.M., & Wills, P.R. (1998). Understanding Maori epistemology: A scientific perspective. In H. Wautischer (Ed.), *Tribal epistemologies: Essays in the philosophy of anthropology.* Sydney: Ashgate, pp. 43-77.

Rose, H. (1994). The two-way street: Reforming science education and transforming masculine science. In J. Solomon & G. Aikenhead (Eds.), *STS education: International perspectives on reform.* New York: Teachers College Press, pp. 155-166.

Ryle, G. (1954). The world of science and the everyday world. In G. Ryle (Ed.), *Dilemmas*. Cambridge: Cambridge University Press, pp. 68-81.

Samovar, L.A., Porter, R.E., & Jain, N.C. (1981). *Understanding intercultural communication*. Belmont, CA: Wadsworth.

Sardar, Z. (1989). *Explorations in Islamic science*. London: Mansell.

Sardar, Z. (1997). Islamic science: The contemporary debate. In H. Selin (Ed.), *Encyclopaedia of the history of science, technology, and medicine in non-western cultures*. Dordrecht, Netherlands, Kluwer Academic Publishers, 455-458.

Scantlebury, K. (1998). An untold story: Gender, constructivism & science education. In W.W. Cobern (Ed.), *Socio-cultural perspectives on science education: An international dialogue*. Dordrecht, Netherlands: Kluwer Academic Publishers, 99-120.

Schilk, J.M., Arewa, E.O., Thomson, B.S., & White, A.L. (1995). How do Native American children view science? *Cognosos, 4* (3), 1-4.

Shelley, R. (1993). *Culture shock*. Portland, Oregon: Graphic Arts Center Publishing Co.

Simonelli, R. (1994). Sustainable science: A look at science through historic eyes and through the eyes of indigenous peoples. *Bulletin of Science, Technology & Society, 14*, 1-12.

Snow, C.P. (1964). *The two cultures*. New York: Menton Books.

Snow, R.E. (1987). Core concepts for science and technology literacy. *Bulletin of Science Technology Society, 7*, 720-729.

Solomon, J. (1983). Learning about energy: How pupils think in two domains. *European Journal of Science Education, 5*, 49-59.

Stairs, A. (1995). Learning processes and teaching roles in Native education: Cultural base and cultural brokerage. In M. Battiste & J. Barman (Eds.), *First Nations education in Canada: The circle unfolds*. Vancouver, Canada: University of British Columbia Press, 139-153.

Sutherland, D.L. (1998). *Aboriginal students' perception of the nature of science: The influence of culture, language and gender*. Unpublished PhD dissertation, University of Nottingham, Nottingham, UK.

Suzuki, D., & Oiwa, K. (1996). *The Japan we never knew*. Toronto, Canada: Stoddart.

Tharp, R. (1989). Psychocultural variables and constraints: Effects on teaching and learning in schools. *American Psychologist, 44*, 349-359.

Traweek, S. (1992). Border crossings: Narrative strategies in science studies and among physicists in Tsukuba science city, Japan. In A. Pickering (Ed.), *Science as practice and culture*. Chicago: University of Chicago Press, 429-465.

Wtson, J. (1968). *The double helix*. New York: Signet.

Wolfe, J., Bechard, C., Cizek, P., & Cole, D. (1992). *Indigenous and Western knowledge and resources management system*. Guelph, Canada: University of Guelph.

Wynne, B. (1991). Knowledge in context. *Science, Technology and Human Values, 16*, 111-121.

Ziman, J. (1984). *An introduction to science studies: The philosophical and social aspects of science and technology*. Cambridge: Cambridge University Press.

J. TURNEY

3. MORE THAN STORY-TELLING–REFLECTING ON POPULAR SCIENCE

Read them [popular science books] closely to appreciate how the authors sustain interest, how they impose order on seemingly disparate facts and events, how they communicate ideas and information without being pedantic, how they inform by telling stories.

Wilford, 1998, p.50

1. THE NATURE OF THE BEAST

Contemplating the contribution of popular science books to contemporary science communication is both exciting and daunting. Exciting because if you happen to like your science packaged in that good old-fashioned commodity the printed book there are more science titles in the bookstores now than ever before. Daunting for the same reason, and for a couple of others too.

One is that this profusion of titles in the basic publishers' category 'popular science' is rather diverse. There is no single, easily defined genre here, rather an area in which many styles and formats co-exist (Jurdant, 1993, Turney, 1999). A more serious difficulty, though, is that it is not obvious where to look for ideas about how to evaluate this growing body of literary work. In 1962 the critic Martin Green declared that 'There is no serious or stringent idea available of what makes a book a worthy example of popularization' (Green, 1964, p.34). Although there are worthwhile critical essays on a number of individual authors (McRae, 1993), it is still true that we lack an effective critical vocabulary for discussing popular science books.

That vocabulary, I think, will have to be quite extensive, and even if I were in a position to do so there would not be space here to develop it in full. But this chapter will offer some hints about the kind of critical work we might expect it to do. As I elaborate them, I have three guidelines in mind. My comments should relate to the particular characteristics of books about science for non-professionals which may set them apart from other non-fiction. They should, if possible, make use of the academic literature in science studies or science communication. And they should offer some grounding for advice to those who wish to write such books, as well as a basis for analysing aspects of those already published.

I will mention three such aspects briefly, then go on to discuss two of them in more detail. The first is the kind of story-telling which goes on in these books. Every successful non-fiction writer will tell you that the way to engage the general reader is to tell a story. Many of these stories are pretty much like stories about anything else - they involve struggles, conflicts or adventures, have heroes and villains, complication and resolution.

S.M. Stocklmayer et al. (eds.), Science Communication in Theory and Practice, 47–62.
© 2001 *Kluwer Academic Publishers. Printed in the Netherlands.*

There is at least one kind of story, though, which is largely confined to the pages of science books–the story of the universe. The historical sciences are obliged to create accounts of change through time. This is perhaps better termed a narrative rather than a story, but it does give these sciences an obvious framework for popular authors to adopt. The fact that the narrative has recently been both extended and filled in with more detail makes the framework even more attractive. One important influence on science as story in the latter half of this century–along with the extraordinary vistas opened up by modern astronomical observation–has thus been the rise of cosmology as a scientific specialty. As Timothy Ferris observes, 'From the perspective of cosmic history, all scientific questions turn ultimately into narratives'. (Ferris, 1997, p.21).

This puts cosmology alongside the other sciences whose business it is to reconstruct a timeline: geology, evolutionary biology and paleoanthropology. One could add developmental biology for completeness' sake. All of these are inevitably in the business of narrative, and the details of the narrative are influenced by theoretical commitments of scientific authors, as Misia Landau, for example, has documented thoroughly for stories of human evolution (Landau, 1991).

It is also easy to observe that a large proportion of recent popular science texts draw on the work of these sciences, often recasting the technical narratives relatively lightly. Martin Eger has characterised this ensemble of texts as a new and important sub-genre of popular science, which collectively relate what he dubs the 'new epic of science' (Eger, 1993). However, while the existence of this evolutionary epic is important for understanding the appeal of many popular science books (Turney, 2001), there is more to popular science than this. And there is more to the art of writing popular science than story-telling. Furthermore, it is difficult to offer general advice about this aspect of the work, as the stories which can be told are heavily constrained by the details of the area of science and by the situation and knowledge of the author.

A second consideration which applies to the area as a whole, and is more specific to popular science, is the treatment of the nature of science. This is of interest for a number of reasons. Every narrative which relates something of what science has found, and how it was found, takes a view on the status of scientific knowledge, sometimes implicit, often explicit. Some understanding of the nature of science is also a component of most definitions of that elusive qualification for modern citizenship, scientific literacy. Finally, it is of interest because you do not have to delve very deep into the mass of popular science books to find that ideas about the nature of science are hotly contested.

This suggests that one aspect of our critical evaluation of any popular science book will be related to the way it treats the nature of science. Aside from whether we happen to agree with the particular version of the nature of science on offer, we will want to know if it is clear what the author wants to convey about this matter, and whether he or she is coherent or consistent about it. By the same token, would-be authors would do well to consider which views are in play in current popular discussion, and to reflect on where their own fit in. All of these reasons add up to a good case for making the nature of science the first topic to get closer attention below.

The second topic is explanation. If there is one thing you have to do as a science writer, it is to explain unfamiliar ideas and phenomena. But it is hard to find any systematic account of how this is done. Advice to authors usually boils down to avoiding equations, minimising jargon, striving for clarity, and using apt metaphors and analogies. It is certainly worth bearing all these in mind. Yet even when you do, it is still hard to define when the job has been done well for the intended audience.

The first of our topics obviously relates directly to academic science studies, just as the popular arguments about the nature of science are a variation on controversies about scientific knowledge which are alive in the academy. For the second, I shall turn to an area of research which science communicators are sometimes prone to ignore, on science teaching. We shall explore whether one set of terms for describing the explanatory performances of a teacher in a science classroom can be applied to the accounts science authors render of similar topics in print.

2. TRUTH, GAME, OR TRUTH GAME: VARIETIES OF SCIENTIFIC METHOD

To get a first perspective on the nature of science, consider the opening of a philosophical work which makes a historical point.

> Once, in those dear dead days, almost, but not quite beyond recall, there was a view of science that commanded widespread popular and academic assent. That view deserves a name. I shall call it Legend.
>
> Legend celebrated science. Depicting the sciences as directed at noble goals, it maintained that those goals have been ever more successfully realized. For explanations of the successes, we need look no further than the exemplary intellectual and moral qualities of the heroes of Legend, the great contributors of the great advances. Legend celebrated scientists, as well as science (Kitcher, 1993).

Philip Kitcher's point, of course, is that the legendary view is now widely disbelieved; that, in the thirty-odd years between the first edition of Kuhn's *Structure of Scientific Revolutions* and his own book, much historical, sociological and philosophical scholarship has called into question the picture of scientists as a band of heroes who, by their command of Scientific Method, wrest truth from Nature's grasp.

Kitcher's own intent is to evaluate the sources of this scepticism and, while acknowledging its force, nevertheless defend a version of the growth of science which provides for rational discussion of scientific progress. I am not going to comment on how well he succeeds. There are much more systematic evaluations of this widely available (Ziman, 2000). I simply want to use him to underline that what once seemed rather straightforward is now contested. In scholarly discussion, the legend is dead - what is at issue is what account of science to put in its place.

Kitcher suggests that the critique worth taking seriously is not that of what he calls 'science bashers' but of those who maintain that Legend offered 'an unreal account of a worthy enterprise'. He also maintains that Legend may still be found in 'textbooks and journalistic expositions'. But thirty years is long enough for news this important to filter out of the academy into the popular realm. And in recent popular science, too, assumptions about the nature of science have become contested terrain

One could take this up by looking at contemporary examples but consider instead someone who has been dead for 100 years, Louis Pasteur. This has the advantage that there is no dispute about the status of the science. But, as we shall see, there is ample scope for disagreement about how Pasteur's claims were established.

For instance, biochemist Frank Ashall in his enthusiastically titled *Remarkable Discoveries!* tells the story of Pasteur's work refuting spontaneous generation, and concludes that;

> Throughout Pasteur's life he encountered opposition to his theories, but he always triumphed over his antagonists. The main reasons for his success are probably his perseverance, conviction of the correctness of his ideas, the brilliance and simplicity of his experiments and the care with which they were performed. In the end, the results of his experiments were self-evident and nobody could argue with them, and they usually turned out to be correct. (Ashall, 1994, p.150)

Just the previous year, though, a pair of sociologists of science published a popular volume with the same publisher, Cambridge University Press, which relates the same episode, but comes to a very different conclusion:

> Pasteur was a great scientist but what he did bore little resemblance to the ideal set out in modern texts of scientific method. It is hard to see how he would have brought about the changes in our ideas of the nature of germs if he had been constrained by the sterile model of behaviour which counts, for many, as the model of scientific method. (Collins and Pinch, 1993, p.9)

Not long after those two books appeared, science historian Gerald Geison published a new full-scale biography of Pasteur. This was generally well-received by the reviewers, with the notable exception of the Nobel prize-winning molecular biologist Max Perutz, who wrote an extended attack on Geison in the *New York Review of Books*. Perutz's intentions were admirably clear:

> Pasteur led a simple family life and devoted all his time to research. To generations of Frenchmen, and to many others, Pasteur has been the image of the selfless seeker after the truth who has been intent on applying his science for the benefit of mankind. In *The Private Science of Louis Pasteur*, Gerald L. Geison, a historian of science, claims to have deconstructed Pasteur, and to have produced 'a fuller, deeper and quite different version of the currently dominant image of the great scientist'. I propose to deconstruct his deconstruction and restore the rightly dominant image. (Perutz, 1995)

He interprets Geison's painstaking historical treatment, drawing on new material from Pasteur's laboratory notebooks, as a salvo in the culture wars, the work of an arch-relativist out to do down a great scientific hero. So according to Perutz, Geison 'insinuates' that Pasteur cheated, 'implies that Pasteur acted dishonestly' and makes the 'accusation' that Pasteur was unethical. He denies that Pasteur could have been guilty of securing a verdict over his opponents in the spontaneous generation controversy at least partly by rhetorical means (Geison's interpretation) because, for Perutz, 'good research needs no rhetoric, only clarity'.

One could follow these arguments through, in Geison's reply to Perutz, or by looking at numerous other popular and scholarly accounts of Pasteur's life and work, but the basic polarity is clear. It is that between realism and constructivism, between seeing scientific knowledge as a true account of an underlying reality, justified by appeal to unequivocal facts, and regarding it as socially, culturally or linguistically constructed, with a much more problematic relation to the real phenomena.

If we want to explore this further, it is important to stress that, as well as being about much more than the image of Pasteur, the range of views about the nature of science is much more complex than this simple opposition suggests. We need to reduce that complexity to organise a discussion of popular science, but not quite as far as this. A useful simple inventory is offered in a British educational researchers' study of young people's images of science. (Driver, Leach, Miller and Scott, 1996). They also needed a way of classifying views of the nature of science, to help interpret students' discussions rather than popular texts, but their list will serve as a starting point.

In reviewing perspectives on the nature of science, they list five main widespread views:

> An *inductive* view of science, in which observations are paramount;
> A *Popperian* view, in which researchers try and falsify conjectures;
> A *Kuhnian* view, in which normal science is punctuated by revolutionary changes of paradigm;
> A *social constructivist* view, in which knowledge is seen as a creation of the relevant scientific community;
> An *instrumentalist* view, in which what 'works' is the principal test of which theories are accepted.

Just as their study applied these illuminatingly to the way school students think about scientific knowledge, I suggest they can help us interpret the epistemological assumptions of popular science texts. Let us try this out on a few examples.

Quantum physics is an area of science which inevitably raises questions about relations between observer and observed, causality and determinism, so books on the subject are likely to treat questions about the nature of the knowledge being described in some detail. Heinz Pagels' *The Cosmic Code - Quantum Physics as the Language of Nature* is a good example. The book was a relatively early contribution to the spate of 1980s volumes about the new particle physics. As the subtitle indicates, it takes an explicit position on the status of quantum mechanics. And it is an elaborate and subtle one. In the foreword, Pagels tells us that 'finally, there is a short third part, *The Cosmic Code*, which describes the nature of physical laws and how physicists find them' (p.12).

Turning to that section, we find a fairly lengthy discussion of the nature of science, which tells us that:

> Looking for natural laws is a creative game physicists play with nature. The obstacles in the game are the limitations of experimental technique and our ignorance, and the goal is finding the physical laws, the internal logic that governs the entire universe. (p. 299)

He suggests that the ultimate laws of nature are not yet known, but that we know what form they will take. They will have a set of characteristics, which he specifies, 'which reflect the relation of our mind to the world it attempts to grasp'.

The scientists are trying to decode the universe.

> The activity of the theoretical physicist is to perceive the internal logic of nature. His interpretations of nature are called theories; they are pictures of the material world made to render it comprehensible.... (p. 305)

On the other hand, 'often the experimentalist discovers a whole new reality' (p.307).

As this suggests, Pagels' view is not easy to classify simply. He has a sophisticated view of the interplay between theory and experiment, is aware that the 'fit' between mathematical statements and reality may be an assumption rather than a discovery, and does not claim that physics has (yet) attained ultimate truth. First and foremost, though, he is a Popperian. For instance, he insists that

> An important feature of modern theory is not that the conclusions are provable but that they must be disprovable... A theory cannot be right in general unless it can be wrong specifically. This epistemic vulnerability lies at the heart of experimental science; indeed the essence of the scientific method is the willingness to place one's ideas in jeopardy. (p. 308-309)

An author like Pagels, writing about well-established science, can tell us what science is like. One writing about contentious science may also feel bound to tell us what it should not be like. An instructive example is Frank Close's *Too Hot to Handle - The Race for Cold Fusion.* (Close, 1992).

This, another physicist's book, is Close's blow-by-blow account of the rise and fall of cold fusion in the mid-1980s. He has an explicit message about the nature of science, and it is a prescriptive one. He tells us at the outset that

> when I set out to write this story I thought that it would primarily show how scientists come up with ideas, design a strategy to test them, carry out the experiments, share experiences and attempt to replicate claimed discoveries, thereby establishing new natural phenomena...

In other words, like Pagels, he expects scientists to work in a Popperian fashion, attempting to refute their conjectures by testing them against reality. However it turned out that the scientists in this story did not keep to these rules:

> but as the months passed I became increasingly aware that there was much more than just science at work here. There were accusations of plagiarism, of cheating, of fabricating data; there was name-calling in public... there were deep politics, paranoia and, by the end of the year, threats of writs... (p. 13)

And at the end of an exhaustive account of the whole cold fusion saga, his conclusion is that

> this is not the way science should proceed, and I hope that readers who have persisted this far will realise that this is not normal science. (p. 349)

He is not referring here to normal science in the Kuhnian sense, but to rational, ethical conduct of science. As the rest of the narrative takes a Popperian view of scientific method, one of Close's major charges against the proponents of cold fusion is that they either deluded themselves, falsified results, or simply suppressed results which didn't fit. If this goes on, then refutation is delayed, and the healthy body of science is contaminated with sub-standard data which is eventually rejected by other scientists working honestly to discover the truth.

This has the virtue of clarity, and also contrasts clearly with some other accounts of cold fusion. Collins and Pinch, for example, in the same volume in which they discuss Pasteur, examine cold fusion and conclude that

> In cold fusion we find science as normal. It is our image of science which needs changing, not the way science is conducted. (Collins and Pinch, 1993, p. 78)

Having pointed to this direct contradiction of Close's conclusions, it is also worth noting that his text, and especially the epilogue added two years after the first edition, does not always bear out his prescriptions. As time passed, not all reputable scientists were convinced by the evidence against cold fusion as rapidly as Close thought they should have been. He is reduced to calling those who continued to argue on behalf of cold fusion, and did not seem interested in the possibility of refutation 'true believers'. His assumptions make it difficult for him to account for why only some scientists found the evidence against cold fusion compelling without accusing the others of bad faith. But there are other views of scientific practice which, without necessarily sacrificing claims to truth, would see this as perfectly rational conduct in a scientific dispute. Here we begin to see how acquaintance with a broader set of views about scientific practice can provide some critical purchase on popular science.

Close writes in defence of a particular view of how scientists ought to conduct themselves, but this is not the only purpose notions of the nature of science can serve in popular texts. For a rather different use, consider a collaboration between a professional physicist and a science writer, Paul Davies and John Gribbin's *The Matter Myth - Beyond Chaos and Complexity* (Davies and Gribbin, 1991). In their preface we find the book is explicitly grounded on a Kuhnian view of science, and on the claim that we are now living through a new Kuhnian revolution across the natural sciences:

> as we approach the end of the twentieth century, science is throwing off the shackles of three centuries of thought in which a particular paradigm - called 'mechanism' - has dominated the world view of scientists. (p. 2)

They claim that we are now heading toward a 'post-mechanistic' paradigm, embracing cosmology, the chemistry of self-organising systems, chaos, quantum mechanics and particle physics. It is unlikely that Kuhn would recognise ideas at this level as fitting his notion of 'paradigm', but what they want to argue is that mechanistic science - which they also toy with calling a myth - only captures part of reality:

> a particular paradigm is neither right nor wrong, but merely reflects a perspective, an
> aspect of reality that may prove more or less fruitful depending on circumstance. (p.
> 3)

They thus adopt a somewhat relativistic view of paradigms, as others have done, though it might be more accurate to say that they make use of the Kuhnian terms to launch a larger claim about overarching world views. There are subsidiary notions about the nature of science, the role of experiments and observation, and so on, in the main body of the text, but I think this rhetorical use of paradigm, as an invitation to the reader to get excited about being part of an intellectual revolution, is the most interesting feature of the book's treatment of science. James Gleick's best-selling *Chaos–Making a New Science* (Gleick, 1988) makes similar use of the Kuhnian distinction between normal and revolutionary science to inflate his claims for the significance of work on chaos, complexity and non-linear systems.

If use of ideas about paradigms, while not necessarily leading to out and out constructivism, usually introduces a relativising account of science, they can also be used to treat ideas conventionally regarded as *outside* science. An especially striking recent example here is the science writer George Johnson's *Fire in the Mind - Science, Faith and the Search for Order* (Johnson, 1995).

This is a book in which epistemological concerns are explicitly taken up at the outset, and figure prominently in the entire text. It examines the competing or complementary belief systems Johnson encounters on a tour of New Mexico - taking in physics at Los Alamos, chaos and complexity theory at the Santa Fe Institute, Catholicism among the Latinos, and the beliefs of the Native Americans still living there. Johnson emphasises that *instrumentally* Western science is the superior system (atom bombs work). But he also argues that all stem from the same underlying human impulse - to lay an ordering grid over the perplexing phenomena of the universe. He thus leans quite a long way toward a constructivist view of theorising. Explicitly, though, he is deliberately declining to choose between constructivism and realism. Johnson identifies the same polarity we found in accounts of Pasteur:

> There are two opposing ways to view the scientific enterprise. Almost all science
> books, popular and unpopular, are written on the assumption that there actually are
> laws of the universe out there, like veins of gold, and that scientists are miners
> extracting the ore. We are presented with an image of adventurous explorers
> uncovering Truth with a capital T. But science can also be seen as a construction, a
> man-made edifice that is historical, not timeless - one of many alternative ways of
> carving up the world. (p. 6)

But he suggests there is no way to choose between them.

> This book...takes an agnostic stance - between the extremes of science as discovery
> and science as construction. In the end, there is no way to know whether science is
> converging on a single truth, the way the universe really is, or simply building artificial
> structures, tools that allow us to predict, to some extent, and to explain and control. (p.
> 6)

This position is not justified with any detailed philosophical arguments, any more than is Ashall's assertion that experiments speak for themselves, but it is a line he follows more or less consistently throughout the book–and he really does try to treat, say, the Los Alamos particle physicists and the Indians dancing to keep the universe in being even-handedly. There are long passages where the descriptions of particular parts of contemporary science are rendered much as they might be in any other popular exposition. But he always returns to the overall question about how we can have confidence in what we think we know.   Johnson's text provides a good example of how an epistemological stance which is still unusual in popular science can provide the inspiration for a successful popular book.

These few examples can only be suggestive, but to my mind they support the idea that the author of a popular science book must pay particular attention to getting their story straight about the nature of science. It will be one of the things which determines the overall coherence of the product. In addition, in order to persuade one's readers of the validity of the position on the nature of science which the book embodies, it will be helpful to be aware of some of the alternatives which are on offer, in popular books as well as in more technical discussions of the production of knowledge. As readers, we want to know what the metascientific message of the work is, and as writers, we need to be clear what message we are giving.

3. DESCRIBING NEW WORLDS: EXPLANATION IN SEARCH OF A THEORY

Whatever stance they take on the nature of science, the chances are that one of the concerns of popular science texts is to explain concepts, experiments and results to readers who are unfamiliar with technical vocabulary or professional technique. This explaining is one of the core attributes of popular science texts and one of the things which distinguishes them from general expository non-fiction.

Some of the difficulties here are experiential. It may be true that science writing takes the reader into new worlds in a way which can be compared with travel writing, but the new worlds of science contain new kinds of entities, as well as existing in realms beyond the normal human senses. They frequently harbour phenomena which contradict intuitions tuned to the middle range of existence. Human beings have direct knowledge only of things of medium size - a few millimetres to a few hundred metres - and which last for a few seconds to a few decades.   They can recover information directly (ignoring the mediation of the sensory organs) only through interaction with light of certain wavelengths, sound of a relatively narrow range of frequencies, and so on. Scientific observation transcends all these limitations.

Then, of course, there are conceptual difficulties. What kinds of things are the scientifically defined entities - atoms or genes - which no-one ever normally sees? What does it mean to conceive of gravity as a distortion of space-time around a massive object? What are virtual particles, electron tunnelling, evolutionarily stable states or hydrogen bonds? And if we want to learn about any of these things, how do we recognise an effective explanation when we see one?

Many popular writers, one suspects, build their explanations by trial and error - seeing what looks right, or sounds intelligible as they work on their text, noting readers' difficulties, and checking out others' explanations of their subjects. A few take a more systematic approach. The physicist Russell Stannard, for instance, has described in detail how he developed the style of explanation he adopts in his series of story books intended to impart the basics of modern physics to young readers (Stannard, 1999).

For a more general account of explanation, though, I want to turn to the educational literature, and to a recent study by John Ogborn and his colleagues. In their book *Explaining Science in the Classroom*, they describe a project which combined linguistic and cognitive science approaches to what happens in science classrooms when teachers are trying to impart understanding of new phenomena (Ogborn, 1996).

Analyzing videos and transcripts of a series of teaching sessions in secondary schools, they develop a theory of explanation based on the use of words, gestures, and objects of diverse kinds to create stories about the kinds of things there are and the kinds of things they can do. The task of explaining is seen as being one of bringing 'entities of science' into being for students. Their account of explanation has four parts: creating differences, constructing entities, transforming knowledge and putting meaning into matter.

Creating differences is their way of summing up the driver for any kind of communication. As they put it,

> in our view the fundamental motor of communication is that there is something known
> to one participant and not – or often *assumed* not to be – known to another.

In science writing one way of translating this is to say that for an explanation to be called for, there must be an acknowledged difference in knowledge. The writer knows why something happens, how it works, or why the accepted explanation is wrong. As we shall see, there may be a prior stage, needed to establish that the subject of the text actually requires explanation at all, a kind of defamiliarisation.

Constructing entities is a matter of populating the explanatory story with things which can behave in a way which accounts for the phenomenon to be explained. Or, as the authors put it

> much of the work of explaining ... concerns the resources out of which explanations
> are later to be constructed. Protagonists have to be described, with what they can do
> and have done to them, before any story which explains a phenomenon can be told.

The more familiar way of describing this is to talk in terms of introducing scientific concepts, but entities is deliberately chosen here as a more general term, covering a very varied set, an 'ontological zoo'. Metaphor and analogy are important here, to help people understand new things in terms of ones which they are already familiar with.

Transforming knowledge means the continuous process of making and remaking ideas, to which the classroom explanation makes one series of contributions. Analogy and metaphor again play key roles here, as does narrative.

Finally, putting meaning into matter is intended to suggest the role of demonstrations, which are supposed to show that the material world really does behave in the ways the ideas and entities which have been brought into play imply.

The authors suggest that these four sets of ideas offer 'the beginning of a new language for thinking about the act of explaining science in the classroom'. But they can readily be adapted to analysing the explanatory work being done outside the classroom, in popular science texts. Let us consider each of their four terms again, in the abstract, before testing this out with some examples.

In text, creating differences will mean trying to establish the differences between what the reader or people in general think is the case, or what is generally taken to be true, and what is the case according to the author. The contrast may be signalled by a range of devices, including imputing views to the reader directly, describing what someone else thinks–where the someone else may be another scientist or a historical actor or someone from another cultur–or outlining a 'commonsense' view.

Elaborating on constructing entities, Ogborn and his colleagues write:

> A scientific explanation needs to invoke protagonists which are not part of common knowledge. Explaining to someone then requires describing the possible protagonists as well as accounting for what they may have done.

Authors have to find ways of introducing the cast of characters they want to use in their explanation - gene, enzyme, atom, antibody, black hole, neuron, magnetic monopole, and so on - and describing their properties.

Transforming knowledge may be a more subtle matter, and harder to locate in texts, but seems most likely to appear in popular texts which have extended treatments of single topics. Sometimes, one is invited to think of some entity one way, by way of introduction, then later on invited to think of it another way, as a substitute, a supplement or an addition to what was said before. This may often be done through telling a historical story.

When they discuss putting meaning into matter Ogborn's group are strictly referring to something which happens outside text. They use this phrase to describe the use of demonstrations as a controlled display of a particular kind. The textual equivalents of this work in popular science books would be *accounts* of experiments or demonstrations. We might also consider that they include descriptions of thought experiments, or various kinds of devices which try and induce us to see things from the point of view of the entities being described: 'if you were an electron/white blood cell/free radical/cosmic ray...'.

Finally all of this work is supposed to be in the service of explanatory stories–these are not stories in the sense of the narratives discussed earlier in this chapter, but accounts which 'tell how something or other comes about'.

So: can we relate these terms to what we find when we examine popular science texts? And are they any help in defining the strengths and weaknesses of particular authors' explanations? Elsewhere I have looked at this in relation to some recent efforts to explain esoteric physics (Turney, 2001b). Here I focus on some examples in the life sciences.

One reason for doing that is it gives an excuse to look the writing of someone who is widely acknowledged to be brilliant at explaining, Richard Dawkins. And if we examine his classic exposition of Darwinian theory *The Blind Watchmaker* (Dawkins, 1986), it is not hard to find examples of each of the stages Ogborn points out.

He begins with a version of creating differences which can often be found in science books. It is not that the reader, or people in general, necessarily have the wrong idea about something, or have learned the wrong solution to some problem–it may be that they do not realise that there *is* a problem. In order for us to attend to an explanation, we must first recognise that there is something which needs explaining. In this case it is the existence of the extraordinary variety of complex living forms.

In the opening paragraph of his preface, Dawkins writes:

> I wrote the book because I was surprised that so many people seemed not only unaware of the elegant and beautiful solution to this deepest of problems but, incredibly, in many cases actually unaware that there was a problem in the first place!

So he is committed from the outset to a two-part strategy. First he must constantly emphasise the wonder and mystery of the existence of complex design in the living world.

> But having built up the mystery, my other main aim is to remove it again by explaining the solution.

In order to outline the workings of Darwinian variation and selection in modern scientific terms, Dawkins has to introduce the essential concept - create entities in Ogborn's terms - starting with the gene. Eventually, he will work up to explanatory stories like the following:

> The protein making machines read the RNA working plans, and turn out protein molecules to their specification. These protein molecules curl up into a particular shape determined by their own amino acid sequence, which in turn is governed by their own DNA code sequence of the gene G. When G. mutates, the change makes a crucial difference to the amino acid sequence normally specified by the gene G, and hence the coiled up shape of the protein molecule.

This is just part of a much more elaborate story about how natural selection might favour an alteration in a gene which affected the behaviour of beavers, by making them hold the wood they use for building dams clear of the water when they carry it to the dam. This, molecular biological part, is chock-full of entities which the reader may not have met before, including proteins, ribosomes ('protein making machines'), and RNA. Each must be introduced, for the story to be intelligible. But they are relatively minor players in the overarching story of the book. The crucial entity is the gene, and its treatment is accordingly more extended, and more elaborate.

Thus, when Dawkins wants to convey an essential attribute of the gene, that it contains a digital code, he does this by likening DNA to a ROM in a computer. Simply stated, this sounds straightforward, and there is indeed a passage in which he first explains the distinction between ROM and RAM in computer memory, and develops the analogy with DNA in detail. But the chapter in which this explanation appears also begins with a powerful piece of scene-setting which I suggest shows how artful prose can put meaning into matter:

> It is raining DNA outside. On the bank of the Oxford canal at the bottom of my garden is a large willow tree, and it is pumping downy seeds into the air... Those fluffy specks are, literally, spreading instructions for making themselves. They are there because their ancestors succeeded in doing the same. It is raining instructions out

there; it's raining programs; it's raining tree-growing, fluff-spreading, algorithms. That is not a metaphor, it is the plain truth. It couldn't be any plainer if it were raining floppy discs. (p. 111)

We might quarrel with Dawkins' assertion that this is no metaphor (I certainly would), but this is a neat textual inversion of the idea of the demonstration. You do not need to contrive for nature to behave in any special way, simply to describe familiar behaviour in new terms.

Dawkins returns to the properties of genes over and over again, so it is not surprising that they also yield the book's best examples of transforming knowledge. In his discussion of development of an individual organism he wishes to shed the strongly deterministic implications of the gene as it has been defined up until then. He does this by an extended comparison of two possible metaphors for the instructions in DNA - as a blueprint or a recipe. The latter is a better way of thinking about the genome, he argues, because

a recipe is not a scale model, not a description of a finished cake, not in any sense a point-for-point representation. It is a set of instructions which, if obeyed in the right order, will result in a cake. (p. 295)

One could find many more examples of all these things in this and in Dawkins' other books, but these suffice to show that he does indeed go through all the stages which Ogborn proposes are part of effective explanation in science. The fact that this is so for a writer whose explanations are so widely admired suggests that there is scope for applying their terms to writing as well as to classroom performance.

In a second example, we can see how this outline of explanation can shed light on a book whose explanatory successes are more mixed, Francis Crick's *The Astonishing Hypothesis* (Crick, 1994).

The book is both a popular exposition, and Crick's call to arms to fellow researchers to begin work on consciousness. He concentrates on the visual system, and this gives him an unusual advantage in terms of demonstration, or of putting meaning into matter. Unlike most authors he is not limited to description of real or imaginary experiments–he can make his points using our own experience of seeing.

So Crick invites the reader to conduct sixteen experiments on his or her own visual system. (pp.26-48). There is a happy match here between medium and message - the pages of the book, reproducing classic visual illusions, are part of the demonstration. If the results of the experiments fit Crick's predictions, he has an especially powerful means to persuade his reader that the visual system is more complicated than he or she thinks. One might also note that there is a wonderful rhetorical paradox here. Crick is explicitly arguing that we should not accept that 'seeing is believing', because almost the exact opposite is true. But he not only shows us why this is so, but repeatedly uses phrases like 'you can easily see', as if vision is unproblematic.

Other aspects of Crick's explanation are more conventional. He creates differences by repeatedly contrasting what he assumes the reader believes with what he argues is actually the case, using phrases like 'many people believe ...' (p.26), 'it is difficult for many people to accept that ...' (p.33), 'what may surprise you is...' (p.24) and, most emphatically, 'it is easy to show that how you think you see things is largely simplistic or, in many cases, plain wrong' (p.24).

In creating entities, Crick is much less careful than Dawkins - or indeed many other writers. In some cases he uses terms like neuron, nerve, and photon without attempting to define them. In others, he defines a term a few pages after it first appears, and then sometimes simply by quoting a dictionary definition. However, when he devotes an entire chapter to a single entity, such as the neuron, he is paying close attention both to creation of entities, such as spike and synapse, and to transforming knowledge.

For example, his first elucidation of nerve conduction says that:

> If the neuron becomes sufficiently excited, it responds ('fires') by sending an electrical pulse - (a spike) - down its output cable - (its axon). (p. 91)

A few pages later, it is necessary to transform this picture, to get a little closer to the actual chemical workings of the neuron:

> The spike in an axon is not like an electric current in a wire. In a metal wire, the current is carried by a cloud of electrons. In a neuron, the electrical effects depend upon charged atoms (ions) that move *in or out* of the axon through molecular gates made of protein, in the insulating membrane of the cell... (p. 96)

In spite of this, as it were textbook example of transforming knowledge, the chapter is one of the least successful in the book. Crick's usual dry humour is entirely absent, and he seems unsure quite how to pitch his exposition. In contrast to his account of visual illusions, the subject matter now works against him (and perhaps against any writer). Many terms from cell biology have to be introduced in a short space, as well as special terms for the unique features of neurons. Then there are different kinds of neuron, and of neuronal signals, to consider.

The whole picture is rather complex, and as it is built up very fast the main portion of the chapter is very dense. The impression of density is reinforced by a sudden flurry of textual quirks. Many words are put in quote marks, but not for any clearly consistent reason. There are frequent bracketed asides, as well as two or three footnotes per page to qualify or elucidate points in the main text. This, along with errors of punctuation like the misplaced comma in the second quotation above, makes the writing seem hasty, as if there is a certain amount of stuff which the author just wants to get out of the way before resuming the more interesting aspects of the exposition. The immediate air of relaxation and assured clarity which resumes as soon as the next chapter, on experimental techniques, gets under way, reinforces this impression.

Although mostly intelligible, the chapter is far from a pleasure to read. It feels like hard work. But as well as recording this impression, we can use Ogborn's vocabulary to diagnose the reason. It is because this author takes much less care, perhaps for reasons of brevity, in creation of entities than Dawkins does. Maybe there are simply more entities to bring into play in a discussion of how the brain works. Or perhaps it is simply the case that Dawkins has learnt to do this work very slowly and carefully as he builds toward his explanatory stories. Either way, it offers more support for the idea that this kind of analysis of explanation can help us identify effective writing.

4 CONCLUSION: DECISIONS, DECISIONS...

I began by endorsing Martin Green's old assertion that we have little idea what makes a book a worthy example of popularisation. This brief review of ideas in just two areas does not make up that deficiency. A broader view of popular science would have to incorporate treatments of the rhetorics in play, the motives of the authors, the selection of topics, and the wider political or cultural effects which such books may have. The diversity of the category is such that this may never be feasible, although we are beginning to see attempts along these lines for some particular topics in popular science, such as Andrew Brown's account of *The Darwin Wars* (Brown, 1999).

However, while the examples I have discussed here are only a beginning, they do show that ways of analysing ideas about the nature of science, and the construction of explanations within the text, may be part of the critical vocabulary we need to bring to bear on popular science. Work with students who have had an opportunity to engage with a wider range of texts also suggests that these ideas have wider applicability within popular science.

Whether they may also be helpful to prospective authors is more problematic. Writing a text like this, part fact-bound, part story-telling, and constantly constrained by hard-to-validate assumptions about what the audience already knows or understands, is an incredibly complex task. Introducing a set of new things to consider may ease the path for some people, inhibit others. But there is one general truth about learning to write which speaks in their favour Any piece of writing is the product of innumerable individual decisions, of numerous different kinds. The effective writer has gone through a long process of becoming conscious of those decisions. But that consciousness requires a way of describing what the various kinds of decision are. What I have highlighted here are some of the decisions which are most important for popular science.

Department of Science and Technology Studies, University College London,Gower Street, London.

5. REFERENCES

Ashall, F. (1994). *Remarkable Discoveries!* Cambridge: Cambridge University Press.
Brown, A. (1999). *The Darwin Wars:How stupid genes became selfish gods.* London: Simon and Schuster.
Close, F. (1990). *Too Hot to Handle: The Story of the Race for Cold Fusion.* London: W.H. Allen.
Collins, H., and Pinch, T. (1993). *The Golem – what everyone should know about science.* Cambridge: Cambridge University Press.
Crick, F. (1994). *The Astonishing Hypothesis: The Scientific Search for the Soul.* London: Simon and Schuster.
Davies, P., and Gribbin, J. (1992). *The Matter Myth: Beyond Chaos and Complexity.* London: Penguin.
Dawkins, R. *The Blind Watchmaker.* London: Longman, 1986.

Driver, R., Leach, J., Millar, R., and Scott, P. (1996). *Young People's Images of Science*. Buckingham: Open University Press.

Eger, M.. (1993). Hermeneutics and the New Epic of Science. In M. McRae (ed.) *The Literature of Science: Perspectives on Popular Scientific Writing*,. University of Georgia Press: Athens, 186-212.

Ferris, T. (1998). *The Whole Shebang: A State of the Universe Report*. London:Weidenfeld and Nicolson.

Gleick, J. (1988). *Chaos: Making a New Science*. London: Sphere.

Green, M. (1964). *Science and the Shabby Curate of Poetry: Essays about the two cultures*. London: Longmans.

Johnson, G. (1996). *Fire in the Mind: Science, faith and the search for order*. London:Viking.

Jurdant, B. (1993). Popularization of science as the autobiography of science, *Public Understanding of Science, 2,* 1993, 365-373.

Kitcher, P. (1993). *The Advancement of Science: Science without legend, objectivity without illusions*. Oxford: Oxford University Press.

Landau, M. (1991). *Narratives of Human Evolution*. New Haven: Yale University Press.

McRae M., (Ed). (1993). *The Literature of Science: Perspectives on Popular Scientific Writing*. Athens:University of Georgia Press.

Ogborn, J. (1996). *Explaining Science in the Classroom*. Open University Press: Milton Keynes.

Pagels, H. (1983). *The Cosmic Code: Quantum Physics as the Language of Nature*. London: Joseph.

Perutz, M. (1995). The pioneer defended. *New York Review of Books, December 21,* 54-58.

Stannard, R. (1999). Einstein for Young People. In E. Scanlon, E. Whitelegg and S. Yates (Eds.) *Communicating Science: Contexts and Channels*. Routledge: London, 134-145.

Turney, J. (1999). The word and the world: engaging with science in print'. In E. Scanlon, E. Whitelegg and S. Yates (Eds.) *Communicating Science: Contexts and Channels*. Routledge: London, 120-133.

Turney, J. (2001a). Valuing the Evolutionary Epic: cosmology, biodiversity and story-telling. *Science as Culture, 10* (2), 241 -263

Turney, J. (2001b - in press). Passing it on - redescribing scientific explanations. In J. Cornwell (Ed.), *Explanation*. Cambridge: Cambridge University Press.

Wilford, J., (1997). Writing Books on Science Topics. In D. Blum and M. Knudson, *A Field Guide for Science Writers*. Oxford: Oxford University Press, 43-50.

Ziman, J. (2000). *Real Science: What it is, what it means*. Oxford: Oxford University Press.

4. TOPICAL PERSPECTIVE AND THE RHETORICAL GROUNDS OF PRACTICAL REASON IN ARGUMENTS ABOUT SCIENCE

1. INTRODUCTION

Only four decades ago, scholars across the disciplines assumed that science is unique among human activities in its rigorous conformity both to empirical criteria for evidential appraisal and to formal rules for testing logical inference. Science, at its best, sought answers to questions of inquiry with recourse to foundations which, purportedly, are impervious to the inquirer's own interests, values, and preferences. Accordingly, the discourses of science, with such firm foundations in nature and in logic, are the best vehicle for bringing forward claims to knowledge that approximated, as far as humanly possible, that which is true.

Today, scholars are more apt to assume that science is constructed within a dynamic complex of social processes permeated with human interests, values, and preferences. The actual practices of scientists consist of myriad layers of decision making and judgment down to its logical and empirical core. The discourses of science, then, are not privileged since they adduce claims to knowledge that lack the special legitimacy once afforded through appeal to purportedly indisputable foundations. Those claims, it turns out, are interested, value-laden, and opinionated, as are those adduced in less epistemologically exalted fields of human endeavor.

Scientific discourse brings forward claims that are far more contingent than heretofore believed and, thus, are open to deliberation, debate, and disagreement. The presence of contingency invites *rhetorical* studies. There is a growing literature on rhetorical dimensions of science that looks to expose the play of contingency behind claims to foundational truths and certainties. In a comprehensive review of that field, Campbell and Benson (1996, p.75) saw consensus in 'the recognition that scientific perception is not immaculate; that scientific method is diverse, social, argumentative, and suasory'. They contend further that 'radical' and 'moderate' projects within the field diverge over what one commentator (Pera and Shea, 1991, p.viii) described as the belief 'that science [does] not provide genuine knowledge and that its methods and its results [are] mere social conventions' (p.75).

Those who adhere to foundationalist positions and those who oppose them still share in common the assumption that 'knowledge' rests on indisputable foundations. Foundationalists assert that point, as they stand against what they perceive as the reduction of knowledge to epistemologically groundless power struggles. Many who hold anti-foundationalist stances also assume that knowledge rests on indisputable foundations, but they instead contend that there are no such foundations

63

S.M. Stocklmayer et al. (eds.), Science Communication in Theory and Practice, 63–81.

in science and question the legitimacy of science's claim to special knowledge status.

There is a third position that does not assume that knowledge requires indisputable standards but instead redirects our efforts from what Dewey (1929) once called a 'quest for certainty' and toward explorations of situated, practical reason. Dewey's concern was how to arrive at 'concrete judgments about ends and means in the regulation of practical behavior.' The range of positions that together constitute this 'third way' do not all share Dewey's pragmatism, of course, but despite important differences they together return us to concerns about practical reason[1] (see notes at end of chapter).

In this essay, I propose to show that behind scientists' formal, published discourses operates a special logic of situated, practical reason. For guidance, I extract from rhetorical theory what I have called a 'topical' perspective for examining the informal features of this practical logic of situated argument in science. I shall first give a synopsis of that perspective. I then illustrate how to use that perspective to probe the operations of practical, rhetorical reason behind Watson and Crick's famous announcement of the DNA double helix. I conclude with a discussion of implications for this practical, rhetorical understanding of the 'logic' of argumentation in science, with attention to the problem of science literacy.

2. TOPICAL PERSPECTIVE AND RHETORICAL LOGIC

A topical perspective focuses on a special kind of situated, practical reason as an alternative to foundational approaches to argumentation. Toulmin (1983, p.398) raised questions relevant to that focus when he sought to redirect our thought about argumentation from foundational emphases on rules of universal validity to practical grounds of situated judgment:

> Are these the *right* (or relevant) arguments to use when dealing with this kind of problem, in this situation?   That is, are they of a kind appropriate to the substantive demands of the problem and situation?

Practical approaches to argumentation raise questions that reveal precisely what foundational approaches conceal: they involve 'particular people, in specific situations, dealing with concrete cases, with different things at stake'.   It thus becomes relevant to ask '*Who* addressed this argument to *whom*, in what *forum*, and using what *examples*' (Toulmin, 1988, p. 339).

Toulmin's questions bring us squarely into contact with rhetorical considerations. Practical reasoning requires attention to the demands on argumentation before situated audiences confronted with specific problems. Thus, we must examine arguments for patterns of judgment that constrain specific instances of practical reasoning. Those patterns are relevant to particular situations, and are not universal; they are substantive, not formal; they are practical, not theoretical.

While the operations of demonstrative logic involve reasoning from within the premises of a particular theory or conceptual framework, arguments that "apply theoretical ideas to practical situations, or which seek to criticise those theories, [must] look outside the theories and so become 'practical' or 'rhetorical' arguments" (Toulmin, 1988, p. 348). Toulmin's point can be generalised to any contested or contestible frame of reference, orientation, or conceptual viewpoint. Indeed, we could extend that point to the reportage of data, which also must evoke the right patterns of situated, audience judgment to warrant their acceptance. Put otherwise, even 'facts' never speak completely for themselves. 'External' rhetorical resources are always required.

Toulmin (1988, p. 347) directs our attention to 'trustworthy generalisations' as among the external resources that are available for adjudicating situated, practical relevance and, thus, brings us to the threshold of rhetorical theory of topical invention[2]. Here, we find some guidance on the operations and standards of situated, practical reason. I have used that theory to frame a perspective for conducting analysis of situated argumentation and judgment about science (Prelli, 1989). Conducting a topical analysis of argumentation reveals how participants (1) frame the specific points at issue, and (2) draw upon often taken-for-granted values and other thematic premises as warrants for accepting particular responses to those issues. Those values and themes, those trustworthy generalisations or *topoi* of argumentation, constrain how audiences adjudicate the situated, practical 'reasonableness' (Prelli, 1989, pp. 113-118) or 'appropriateness' (Waddell, 1990, pp. 393-395) of proffered claims[3].

Legitimating grounds that scientists share *as* scientists comprise a fairly consistent set of principles, presumptions, and premises. For instance, a fundamental principle governing scientific argumentation is that reasonable claim making involves maintaining or extending comprehension of some part of the natural world that concerns a specific community of interest. Moreover, it is widely assumed that claims advanced have emerged from prescribed methods and procedures for solving the kinds of technical problems that otherwise would have obscured that community's fuller comprehension (Prelli, 1989, pp. 120-25). Potentially reasonable arguments about science, then, advance claims that identify or modify specific problems that scientists perceive as relevant to their better comprehension of the natural order and do so through accepted procedures and methods.

Though the specific problems chosen and their formulation vary considerably, we still can make generalisations about the *kinds* of problems scientists address whenever they make and evaluate arguments about science (Prelli, 1989, pp.144-158). Scientists identify and address ambiguities or defects in evidence and data, in prevailing theories and interpretations, in received methods and procedures, and in the significance of proffered claims for the research community. Respectively, these recurrent kinds of problems or ambiguities are *evidential, interpretive, methodological,* and *evaluative.* All are 'scientific' problems, in so far as they directly or indirectly impede the efforts of a research community to maintain or extend its comprehension of natural order.

Classical rhetorical theory asserted that problems or ambiguities giving rise to debate can be framed as distinctive kinds of issues. Stripped of circumstantial details, specific issues about any problem or ambiguity are finite and reducible in structure to four fundamental kinds. *Conjectural* issues turn on whether some thing does or does not exist. *Definitional* questions turn on what constitutes the meaning of admitted things. *Qualitative* issues involve questions of how admitted and defined things apply within a particular mix of circumstances. *Translative* issues revolve around which procedures or actions are best for resolving situated problems. These four kinds of issue specify the different ways that scientists frame evidential, interpretive, methodological, and evaluative ambiguities or problems.[4]

Toulmin's 'trustworthy generalisations' resemble what were called in classical rhetorical theory *topoi* or *loci*; they are 'topics', 'regions', or 'places' where one can turn to find premises for potentially relevant lines of argument. What are the *topoi* of argumentation about science?[5] They are a veritable checklist of topics normally taken up when teaching the practical logic of scientific inquiry, if not the techniques and procedures usually glossed behind the simple rubric, 'scientific methods.' An inventory includes empirical accuracy, precision, and consistency with received knowledge. *Topoi* specially associated with theoretical claims include explanatory power, predictive power, scope or generalisability. Those and other themes and values are generally available and potentially relevant resources for argumentation that scientists can use to legitimate their specific claims as reasonable responses to specific points at issue.[6]

This partial summary of the general rhetorical grounds of practical reason shows that argument about science is neither illogical nor nonlogical; it is grounded in communally shared and conventionally understood standards of scientific reasonableness. Of course, the operations of that topical, rhetorical logic always are specific to situation; accordingly, any inspection of that logic requires attention to the practical demands and particular circumstances that generated the arguments under examination.

In the next section I illustrate the operation of topical, rhetorical logic through inspection of the practical reasoning behind Watson and Crick's classic DNA double helix announcement.

3. RHETORICAL LOGIC AND THE DNA DOUBLE HELIX

Watson and Crick's famous DNA double helix announcement (1953) initiated one of the greatest advances in twentieth century biology, furnishing what Kuhn would have defined as an exemplary problem-solution or paradigm for that field (1970). The full context of the path to DNA's three-dimensional structure is reconstructed in careful historical works about the event (Judson, 1979; Olby, 1974).
Also available are autobiographical accounts from leading participants (Chargaff, 1978; Crick, 1988; Watson, 1968). Rhetoricians found the announcement irresistible as a test case for the claim that the discourses of science are indelibly

rhetorical (Bazerman, 1981; Fisher, 1994; Gross, 1990; Halloran, 1984; Miller, 1992; Prelli, 1989).

I contend that the significant rhetorical features of the announcement are displayed prominently through its situated adaptation of an informal, topical logic. The grounds of that logic constrain (1) how issues are framed to focus points for audience adjudication and, (2) how claims are warranted as 'reasonable' or 'appropriate' responses to those issues. The announcement works to constrain audience appraisal of its proffered claims through often implied evocation of technical values and other thematic premises that scientific communities evolved as grounds for assessing 'the scientifically reasonable' during situated, practical problem solving.[7]

Much was already known about the DNA molecule when Watson and Crick started working earnestly on the problem. Its chemical composition was known since the 1930s; it consisted of phosphate and sugar groups (with attached bases). Avery (with co-workers MacLeod and McCarthy) had intimated as early as 1944 that DNA was associated with genetic material. Hershey and Chase made that point more boldly in 1952. Watson and Crick also had data about the dimensions of the molecule from published and unpublished experimental work. The central ambiguity was how all this information could be integrated within a coherent structure for the molecule.

Crick and Watson worked to solve DNA's structure at the Cavendish Laboratory at Cambridge. They were not alone in this effort. Linus Pauling was working on the problem at the California Institute of Technology and, prior to Watson and Crick's announcement, had published (with Robert Corey) a proposed structure of his own. Maurice Wilkins and Rosalind Franklin were at work on the problem at King's College in London.

Watson and Crick's article frames the problem of DNA's structure primarily as an *interpretive* problem. They set out two major claims in the opening paragraph:

> We wish to suggest a structure for the salt of Deoxyribose Nucleic Acid (D.N.A.). This structure has novel features which are of considerable biological interest' (1953a).[8]

The first claim is developed extensively through specific description and applications of their proposed model. As the piece unfolds, we see how the model integrates structural features and dimensions meaningfully within a single configuration. We also see that they did not approach the problem as primarily evidential and, thus, as requiring exacting experimental investigation before adducing their proposed solution. Experimental data were necessary but insufficient means to persuasion in this case.[9] The second major claim turns out to be *evaluative.* Once the model is established, Watson and Crick suggest with considerable brevity precisely what makes the proposed structure 'of considerable biological interest' and, thereby, establish a claim for the theoretical solution's significance.

Watson and Crick's claim to possess an answer to the structural problem logically presupposes that no meaningful theoretical structure existed. They had to argue that point since two models were already extant, including one published by the Nobel Prize winner Linus Pauling (Pauling and Corey, 1953a; Pauling and Corey, 1953b).[10] Accordingly, the next two paragraphs address a *conjectural* issue about interpretation: 'Is there a reliable model for DNA's structure?'

Pauling and Corey's model was a triple helix; it consisted of three sugar-phosphate chains. The phosphate groups faced the molecule's interior axis and the sugar groups, with attached nitrogenous bases, were located at the molecule's exterior. Watson and Crick made two refutative claims about this model. One is that Pauling and Corey had misinterpreted X-ray evidence when they formulated their model: 'We believe the material which gives X-ray diagrams is the salt, not the free acid.' As a result, the model did not show how the three strands were held together: 'Without the acidic hydrogen atoms it is not clear what forces would hold the structure together, especially as the negatively charged phosphates near the axis will repel each other.'

The other claim is that the model violated some 'van der Waals distances', or the distances, as Judson (1979, p.80) expressed it, where atoms 'begin to get in each other's way.' Pauling and Corey had packed the atoms too tightly within the core of the structure.

These claims were warranted with two intertwined valuational standards for judging the 'reasonableness' of claims as science. Pauling and Corey misinterpreted experimental evidence, so their model was *empirically inaccurate*. The model also violated known relationships among atoms and molecules, so it was *inconsistent* with received chemical theory. Empirical accuracy and consistency with established knowledge are two conventional criteria from which 'good reasons' often are forged when particular claims are assessed for their reasonableness as science.

Watson and Crick then adduced one refutative claim against the second proposed three-chain model. Fraser's model, they asserted, is 'rather ill-defined, and for this reason we shall not comment on it'. Fraser's model thus lacked sufficient *precision*. As Fisher (1994, p. 27) put it, 'precision is a value and lack of it is sufficient reason, a good reason, for rejecting ideas that are supposed to be scientifically sound.' Precision, then, is another valuational standard for assessing the reasonableness of claims.

Watson and Crick did not want to mention Fraser at all. As Watson (1968, p.139) put it, Fraser had three groups of bases, many of which were in the wrong tautomeric forms. He and Crick thought this 'idea did not seem worth resurrecting only to be quickly buried'. Wilkins and Franklin insisted that Watson and Crick acknowledge Fraser, who was a member of their lab, because he had considered hydrogen-bonded bases before they did. This brief paragraph, then, is the result of a compromise about *priority* in making claims relevant to the ultimate solution.

Through their refutative stands on the conjectural issue, Watson and Crick opened a conceptual space for inserting their own proposed model into the published literature: 'We wish to put forward a radically different structure for the salt of deoxyribose nucleic acid.' The next four paragraphs describe their proposed model in sufficient detail to address the *definitional* issue about interpretation: 'What is the model of DNA's structure?'

Paragraph four renders the double helix reasonable by indicating throughout how its structural features fit within the framework of existing knowledge. The model consists of *two* helical chains. Watson and Crick 'made the usual chemical assumptions' about their composition: they consist of sugar and phosphate groups with linkages on number 5 and number 3 carbons of the sugar groups. Both chains are 'right-handed' helices related by a dyad perpendicular to the structure's axis. Due to the dyad, 'the sequences of atoms in the two chains run in *opposite* directions' (emphasis added). (In other words, the sequence of atoms in one chain matches the other in reverse.) The structure 'loosely resembles' Furberg's model insofar as the bases attached to the sugar groups are located inside the structure and the phosphate groups are on the outside. Sugar and adjacent atoms are configured "close to Furberg's 'standard configuration'.[11]

The paragraph continues with controlling dimensions gleaned from X-ray patterns: discrete 'residues' on each chain repeated every 3.4Å; the entire structure repeated every ten residues (assuming a 36 degree rotation from one residue to the next) and, thus, measured 34Å; the distance of the external phosphorous atoms to the vertical fibre axis was 1Å. To these data the point is added that the negatively charged phosphates are positioned to receive cations at the structure's exterior (which invites comparison with Pauling and Corey's mistaken interior placement of the phosphates where they were likely to repel each other).

Watson and Crick's description in paragraph four warrants the judgment that their model possessed the technical values that Pauling and Corey's model lacked. The double helix was at least *adequate empirically* and clearly was *consistent* with received chemical theory. In addition, one might infer that the description of the atom sequences of the two chains running in opposite directions possessed considerable *originality*; it was a novel insight that emerged against the backdrop of received knowledge. Some degree of empirical accuracy (i.e. adequacy), consistency with received knowledge, and originality or novelty in thought are among the standard *topoi* that scientists use to invent 'good reasons' for warranting specific claims as reasonable science.

The fifth paragraph asserts that the structure has high water content and will become 'more compact' with less water. The point there is that the structure, as described, could accommodate experimental evidence about DNA in both its A- and B-forms. This point is important because of Astbury's previously published X-ray data. Astbury reported that the entire structure repeated every 28Å, but failed to distinguish A- or 'wet' and B- or 'crystalline' (i.e. dry) forms. Franklin had data that showed DNA fibers, when wet, expanded in length from 28Å to 34Å. This paragraph removed a source of possible confusion about two measures that otherwise could have raised ambiguities about the model's empirical support.[12]

Watson and Crick completed their response to the definitional issue in paragraphs six through eight when they described the hydrogen-bonded base-pairing mechanism that holds the two chains together. That description follows:

> The novel feature of the structure is the manner in which the two chains are held together by the purine and pyrimidine bases. The planes of the bases are perpendicular to the fibre axis. They are joined together in pairs, a single base from one chain being hydrogen-bonded to a single base from the other chain, so that the two lie side by side with identical z-coordinates. One of the pair must be a purine and the other a pyrimidine for bonding to occur. The hydrogen bonds are made as follows: purine position 1 to pyrimidine position 1; purine position 6 to pyrimidine position 6. If it is assumed that the bases only occur in the structure in the most plausible tautomeric forms (that is, with the keto rather than the enol configurations) it is found that only specific pairs of bases can bond together. These pairs are: adenine (purine) with thymine (pyrimidine), and guanine (purine) with cytosine (pyrimidine).
> In other words, if an adenine forms one member of a pair, on either chain, then on these assumptions the other member must be thymine; similarly for guanine and cytosine. The sequence of bases on a single chain does not appear to be restricted in any way. However, if only specific pairs of bases can be formed, it follows that if the sequence of bases on one chain is given, then the sequence on the other chain is automatically determined.

This description evokes two interwoven thematic premises that warrant its reasonableness. One is explicit: the complementary base pairing mechanism strengthens the model's claim to *originality* and novel insight. The other, Halloran (1984, p. 73) contends, is an implicit appeal to scientists' sense of *elegance*. The structure's symmetry and balance appeal to the aesthetic. With two chains running in opposite directions, one chain's bases complement those on the other chain. Hydrogen bonding joins purine and pyrimidine bases which, when in 'the most plausible tautomeric forms', allow only specific kinds of pairings to occur: adenine (purine) with thymine (pyrimidine) and guanine (purine) with cytosine (pyrimidine).

The repeated aesthetic judgments of scientists furnish evidence for Halloran's claim that elegance is one of the announcement's most important means to persuasion. Chargaff (1974, p. 778) later called the model an 'aesthetically pleasing solution'. Watson (1968, p. 131) recounted his feeling that 'a structure this pretty just had to exist'. Crick (1974, p.768) believed the ultimate impact of his and Watson's discovery had much to do with the 'intrinsic beauty of the DNA double helix.'

At this point in the announcement, Watson and Crick addressed the *qualitative* issue about interpretation: which of the model's applications are most meaningful? In the ninth paragraph, they contend that the base-pairing mechanism fit important experimental data: 'It has been found experimentally that the ratio of the amounts of adenine to thymine, and the ratio of guanine to cytosine, are always close to unity for deoxyribose nucleic acid'. These are the well established Chargaff ratios. Readers could infer that the base-pairing mechanism fit those ratios and, thus, is an empirically adequate feature of the model. More importantly, they could see the mechanism as possessing the power to explain those ratios; the ratios were inevitable consequences of the mechanism (Halloran, 1984, p. 73). The particular structure was reasonable as science because it possessed at one stroke *empirical adequacy* and *explanatory power*.

There is evidence that scientists would adjudicate the application of the model to the ratios in these ways. Crick (quoted in Judson, 1979, p. 143) saw *explanatory power* as an important feature of the mechanism. He recalled the meeting with Chargaff when he experienced that insight: 'I suddenly realized, by God, if you have complementary replication, you can *expect* to get one-to-one ratios.' Chargaff (1974, p. 778) later pointed to the model's *empirical adequacy* when he wrote that it was 'the most plausible inference from the base-pairing regularities earlier discovered by us in many DNA preparations'. We can surmise that many readers would tacitly furnish similar warranting premises when reading paragraph nine.

In the tenth paragraph, Watson and Crick continue to address the qualitative issue, but this time indicate where the model does not reasonably apply. They write, 'it is probably impossible to build this structure with a ribose sugar in place of the deoxyribose, as the extra oxygen atom would make too close a van der Waals contact.' Application to the other nucleic acid, RNA, is not reasonable because it is *inconsistent* with received chemical knowledge about distances at which atoms get in each other's way.

Watson and Crick complete their response to the qualitative issue about interpretation in the eleventh paragraph with a carefully crafted argument that combined previously published experimental data with data reported in the companion papers in that *Nature* issue. They strengthen their model's *empirical adequacy* as a solution to the structural problem of DNA. They tell readers that previously published experimental data are 'insufficient for a rigorous test of our structure.' So, the model 'is roughly compatible with the experimental data, but it must be regarded as unproved until it has been checked against more exact results.' As it turns out, more exact results are immediately forthcoming in the experimental papers published together with Watson and Crick's announcement:

> Some of these are given in the following communications. We were not aware of the details of the results presented there when we devised our structure, which rests mainly though not entirely on published experimental data and stereochemical arguments.

Following Watson and Crick's paper was an article by Wilkins, Stokes, and Wilson entitled, '*Molecular Structure of Deoxypentose Nucleic Acid.*' Next came one by Franklin and Gosling, '*Molecular Configuration in Sodium Thymonucleate.*' Both furnished X-ray evidence that strengthened Watson and Crick's argument that their proposed model was adequate empirically. Indeed, Franklin and Gosling's piece contained an X-ray photo of DNA in its B-form, and both papers contained important confirming data on the molecule's dimensions.

Wilkins had suggested that he (and his co-workers) and Franklin and Gosling publish their data along with Watson and Crick's model (Judson, 1979, p.178). Watson and Crick could then strengthen the claim for their model's empirical adequacy beyond referring to previously published data. Their careful wording created the impression that data reported in the companion pieces were, to some extent, independently confirming their model while conserving claims for originality in its construction. Consider again the statements (italics added) in paragraph eleven:

> *We were not aware of the details* of the results presented there [in the two articles] when we
> devised our structure, which rests mainly *though not entirely* on published experimental data and
> stereochemical arguments.'

And, in the fourteenth and final paragraph:

> We also have been stimulated by a knowledge of the *general nature* of unpublished experimental
> results and ideas Dr. M. H. F. Wilkins, Dr. R. E. Franklin, and their co-workers at King's
> College, London.[13]

The wording of the argument premised on empirical adequacy and the manner of
crediting other investigators reflect that (1) empirical adequacy is an important
structure of thought for scientific claim-making, and (2) after their article was given
its basic form, Watson and Crick continued to adjust it rhetorically as the
circumstances of their situation changed. This appeal completed their qualitative
stand on the model's situated applications.

Recall the second major claim stated in the opening paragraph about the
proposed model's novel features of 'considerable biological interest'. Paragraph
twelve picks up that claim and addresses a *conjectural* issue about the model's
evaluation: is the double helical model significant as science? No other paragraph in
the paper received so much attention as that one sentence paragraph asserting that
the base-pairing structure also accounts for the molecule's biological function:

> It has not escaped our notice that the specific pairing we have postulated immediately suggests a
> possible copying mechanism for the genetic material.

With the history of the twelfth paragraph available to us, we can see it as the
product of Watson and Crick's deliberation about rhetorical strategy. According to
Watson, Crick originally wanted to expand discussion of genetic implications, but
'finally saw the point to a short remark.' According to Crick, Watson did not want
even to make the claim about replication (let alone extend discussion) for fear that
the structure might prove empirically inaccurate. What ultimately was a very
important point is made with exceptional brevity because the two scientists had to
come up with a verbal compromise that was the product of their differing notions
about what they could actually claim and of what the scientific community would
accept. Crick (1974, p.766) explains how the compromise transpired:

> This [paragraph twelve] has been described as 'coy', a word few would normally associate with
> either of the two authors, at least in their scientific work. In fact it was a compromise, reflecting
> a difference of opinion. I was keen that the paper should discuss genetic implications. Watson
> was against it. He suffered from periodic fears that the structure might be wrong and that he had
> made an ass of himself. I yielded to his point of view but insisted that something be put in the
> paper, otherwise someone else would certainly write to make the suggestion, assuming we had
> been too blind to see it. In short, it was a claim to priority.

Watson and Crick's after-the-fact accounts allow us to see that their rhetorical
difficulties were of two completely intertwined types: (1) what they, as a team, took
as sufficiently reasonable to present publicly in their strategic situation, and (2) what
their readers would think of their credibility, or ethos, as scientists. One could

hardly have a clearer example of how inseparable scientific knowledge and rhetorical adaptation of claims can become when scientists make discourse.[14]

We can infer that readers would grasp readily the powerful point that the molecule's structure simultaneously accounted for its biological function. That insight resolved any ambiguities about the model's significance because its *fruitfulness* is immediately clear; it opened important theoretical and experimental work within the field. As events turned out, the fruitfulness rhetorically implied became a reality, with a spate of important theoretical and experimental work following the DNA double helix into publication.[15]

Failure to develop the theme of fruitfulness might seem an oversight, but more importantly from my vantage point is that its brief development resulted from a *rhetorical* decision. What if Watson allowed Crick to elaborate on the genetic implications of the model? The model's fruitfulness would then become a more important line of argument, but that argument's development would create more opportunities for readers to locate problems with the proposed structure. Fruitfulness was not developed more fully because of a disagreement about rhetorical strategy, but the compromise that resulted yielded three important advantages. The brevity of the statement (1) established their claim to priority for the replication mechanism; (2) indicated the model's immediate fruitfulness by suggesting that the structural solution also accounted for the molecule's biological function–replicating genetic material; and (3) allowed them to avoid complex problems concerning precisely how the chains separate and unwind during replication or how the base sequences act like a genetic code.[16]

Watson and Crick did what scientists must do when addressing claims to their peers. They focused their paper on a problem whose modification was of interest to researchers: What is the structure of DNA? In this case, they framed that problem from an *interpretive* vantage, so that they could adduce a primarily theoretical solution. They addressed the problem with claims that responded to conjectural, definitional, and qualitative obstacles which, if left unaddressed, could block endorsement of their proposed solution. Finally, they warranted those claims with valuational structures of thought that furnished reasonable grounds for endorsement. Their article in *Nature* was compelling because by habit, reason, strategy, or all three, the two scientists argued that their model for DNA promised *empirical adequacy* and possessed qualities of *originality, consistency, explanatory power, elegance,* and immediate *fruitfulness.* These are among the standard criteria scientists evoke during argumentation about scientific matters.

4. RHETORICAL LOGIC AND THE PROBLEM OF SCIENCE LITERACY

Watson and Crick hit upon useful persuasive means to gain endorsement of their model. The influence of their article was not due solely to its bold ideas. Astute rhetorical decisions entered in, several times very consciously. Whatever else 'science' might be, it also is argumentation about situated problems that confront specific audiences who must adjudicate the reasonableness of claims addressed to those problems. Those arguments, I have contended, are constrained by the shared grounds of a rhetorical, topical logic. Some implications follow from this position.

First, topical analysis brings into view the rhetorical operations of situated practical reason. It shows how rhetorical theory can help furnish an understanding of the logic of scientific argumentation that differs from reliance on an idealised analytic logic, on the one hand, and from complete skepticism about there being any actual logic at all, on the other. Instead, we see at work a practical, rhetorical logic of argumentation.

Aristotle directed attention to the operations of this logic long ago when he distinguished analytics from rhetoric and dialectic precisely because where audiences *participate in* creation of premises or conclusions and *extrapolate from* examples, the systems of validation used in analytics simply fail to apply. What Aristotle did not see so clearly is that science, too, involves argumentation for and within a community that *participates in* and *judges* the reasonableness of claims. Accordingly, the logic of scientific argumentation involves exercise of situated reason about recurrent kinds of issues based upon communally 'authorised', warranting standards - standards that militate against purely idiosyncratic or arbitrary participation and judgment. This is the 'topical,' rhetorical logic of scientific argumentation.

The values and themes extracted from Watson and Crick's announcement illustrate some of the warrants that scientists evoke to advance or evaluate claims. Scientists should recognise these often taken-for-granted and conventional patterns of thought as part of their mental equipment acquired from training in scientific thinking and associated methodologies, regardless of specialty fields. For instance, *precision* and *accuracy* are potentially relevant lines of thought for situated deliberations about issues ranging from the chemical composition of proteins to the trajectories of satellite orbits. As *topoi*, they designate general patterns of thought and thus fail to capture specific technical doctrines and information whose mastery would distinguish, say, molecular biologists from rocket scientists. However, without those definite patterns of valuational thought one literally cannot 'talk or write science.' They are among the range of possible 'common grounds' for situated deliberation about what constitutes 'the scientifically reasonable' in particular cases, regardless of specialty area.

This topical understanding of argumentation and deliberation can help demystify what scientists do when they practice their craft. This is especially important in our age. Journalistic accounts conceal behind dramatic portrayals of 'discoveries' or technological wizardry the concrete practical processes involved in actually doing science (Nelkin, 1987; Burnham,1987). As a result, science becomes mysterious and is thought beyond the reach of ordinary comprehension. As one critic of modern culture put it, science then becomes a 'god term'; it designates the ultimate source of modernist meaning that attracts irrational allegiance and commands irrational respect (Weaver, 1985, pp. 211-218). Of course, for many the god term of 'science' has lost much of its evocative power. Consider that in every other news story in modern days, the public is offered the testimony of scientists as at least tentatively 'settling,' only to learn the next day that other 'experts' furnish contrary conclusions. What results is confusion at best, and cynicism about all 'expert' authority at worst.

Worshipful appreciation and cynical frustration are two sides of the same coin. Both assume science is or should be rooted in indisputable standards. In either case, disputed claims are taken as evidence that something other than pursuit of 'real' knowledge has taken place. We have here the public parallel to the divergent positions of scholarly commentators on science foundations mentioned earlier. Should the public understand that science is *unfolding argumentation* - whatever else it may be - they could avoid both consequences of that assumption. Fostering that understanding requires that we reconsider the problem of science illiteracy as it usually is framed.

Science literacy efforts are launched both to rekindle flagging public appreciation for science and to remedy public deficiencies in technical understanding. In Great Britain, the Wolfendale Report wants the public to comprehend 'scientific concepts, terms and issues,' and to appreciate science as vital to national life (UK, 1995, Section 1.8). In the United States (1994), the Clinton Report stated that scientific and technical literacy are required so that citizens can both make informed decisions and acquire appreciation of science.

Prospects for the success of those efforts seem dim. Some have argued that the goal of widespread technical literacy is particularly Quixotic (Shamos, 1995). That position is strengthened by the public's weak grasp of the ideas and concepts that frame conventional *political* practices, let alone scientific practices. Consider Lasch's (1995, p.162) portrayal of the American citizenry:

> Millions of Americans cannot begin to tell you what is in the Bill of Rights, what Congress does, what the Constitution says about the powers of the Presidency, how the party system emerged or how it operates. A sizeable majority, according to a recent survey, believe that Israel is an Arab nation.

If the American public cannot grasp those essentials, why should we expect that science literacy campaigns could elevate their comprehension of arcane technical nomenclatures and complex principles of science?

One of the ironies of our 'age of information' is the absence of widespread public understanding of the substance of politics. For that reason alone, we at least should question technical solutions that variously aim to elevate public understandings through imparting yet more information. We also should explore alternatives to the usual technical solutions.

One alternative to standard technical solutions is a 'rhetorical' approach to science education. That approach assumes that if the public participated in political debates the requirements of argumentative give-and-take would furnish the impetus for seeking out and discriminating relevant from irrelevant information. As Lasch (1995, p.163) put it, public debate itself is educational:

> We do not know what we need to know until we ask the right questions and we can identify the right questions only by subjecting our own ideas about the world to the test of public controversy. Information, usually seen as the precondition of debate, is better understood as its

by-product. When we get into arguments that focus and fully engage our attention, we become
avid seekers of relevant information. Otherwise we take in information passively - if we take it
in at all.

By implication, then, the remedy for enhancing public understanding of science is
the same remedy as that required to restore a vital political public. A rhetorical or
debate approach to science education in the schools, colleges, and universities would
require no less than an understanding of science as practical argumentation, in both
technical and public forums.

That sets a difficult but central task for those concerned with promoting better
public understanding of science.

A topical perspective is well suited to that task. Topical analysis reveals the
issues, values, and themes that scientists recognise as relevant to their deliberations,
but does so at a level of generality accessible to nonscientists. Students in any field,
scientific or other, can be taught to identify those patterns of thought without
acquiring sophisticated levels of technical proficiency within a particular specialty
area. They can, for instance, grasp the salient issues, warranting values, and
thematic premises that structure the practical, rhetorical 'logic' of Watson and
Crick's double-helix announcement and can do so without much technical
information about tautomeric forms, van der Waals distances, and cations.

A rhetorical approach to science education would aim to increase student
comprehension of science as constrained argumentation rather than conduct literacy
campaigns designed to enhance their technical virtuosity. That understanding
becomes all the more crucial whenever science impinges on matters of public policy
or public interest. Students need to learn how to acquire technical information, but
that learning is best experienced within the context of studying alternative positions
on crucial public issues. When confronted with well chosen case studies in the
college or university classroom, students can use a topical approach to identify,
discriminate, and evaluate comparatively the range of values that compete for public
allegiance - whether values of science or other values. Put otherwise, students could
learn the alternative grounds and rhetorical operations of situated practical reason as
they confront important public problems.

Consider some examples. A study of public deliberations about potentially
hazardous research conducted within a community reveals alternative strategies for
framing issues and adapting modes of appeal to specific, situated audiences who are
charged with representing the public's vital interests (Waddell, 1990). A study of
argumentation about increased research funding for new technologies and about the
extension of their use can bring into view a plurality of values that would otherwise
remain concealed, as was the case with reproductive technologies that promised the
ability to 'pre-select' the sex of children (Stearney, 1996). A study of how social
science researchers argue about methods can also show how they build into their
procedures preferred political and moral assumptions which, in turn, help shape the
'facts' that emerge statistically from their studies of politically volatile public
problems, such as family violence (Prelli, 1996). All such cases raise questions
about which from among a plurality of possible values should guide public

deliberations. And as thoughtful students discover the range of options available, they are better prepared to discriminate where 'expert' authority ends and political exhortation begins.

A rhetorical approach to science education should at least augment literacy campaigns that aim explicitly at fostering public 'appreciation' for science - an aim that is more feasible than the quest for widespread technical literacy. Shamos (1995, pp.197- 202) discussed Edward Teller's position that nonscientists are better educated about science from 'appreciation' courses that parallel similar courses in music or art than from technical training *per se*. Shamos (1995, p.201) elaborated on the objectives of this overall approach as follows:

> One objective of general education in science, perhaps the most important one in the eyes of the scientific community, must be to encourage an appreciative audience, one that at least understands how and why so much needs to be spent on science and technology, even apart from military requirements, to keep pace with the developing world. Another objective, probably more important to society at large, must be to help the student, and society generally, feel more comfortable with new developments in science and technology. They need not so much to understand the details but to recognize the benefits -- and the possible risks.

Science appreciation as a goal of literacy campaigns comes with its own risks. For example, Shamos seems to assume that the interests of science and technology have special claim to public treasuries and, aside from acknowledging possible risks, views the interests of society as primarily that of comfortable adaptation to changing circumstances shaped by scientific and technological advance. Those assumptions beg important questions. Should we assume from the start that competing claims to finite public resources are of less importance than are those of science and technology? Should we assume that the public must adapt to technological changes that transform the circumstances of their lives? Or should the public exercise some influence, if not control, over those changes? In any given deliberation about matters scientific and technological, whose interests are advanced and whose are diminished, and for which purportedly 'public' goods? It would seem that those who purportedly speak for the interests of science and technology are no less obligated than others to meet the demands of public deliberation and persuasion; literacy campaigns should not purchase public 'appreciation' of science at the price of foreclosing those obligations.

'Appreciation' of science as *a* public good is a worthy objective, but science education becomes propaganda to the extent that it promotes uncritical acceptance of science as *the* public good.

A rhetorical approach to science education necessarily would foster a *critical* appreciation of science. Understanding the rhetorical logic of situated practical reason that operates in discourses about science neither denigrates nor valorises science. As students learn to distinguish values that warrant technical arguments about science from those used to warrant the acceptability of other purported public goods they become better equipped to adjudicate public issues for themselves. They can, for instance, better detect the lures of scientism and its pursuit of political aims and claims concealed behind technical arguments about science. Or, they can better resist the appeals of pseudo-science and popular superstitions masquerading as science. Above all, students are prepared to act as citizens when called upon to adjudicate the plurality of values, or alternative conceptions of the good, that are put into play during public deliberations involving matters of science. Any 'appreciation' of science worthy of the name would be founded upon such discriminating, situated judgments about the best among alternatively argued public goods. And helping students render such judgments sets an important goal for a rhetorical approach to science education.

Department of Communication, University of New Hampshire, USA

5. NOTES

1. For an excellent overview of major positions that coalesce around a return in philosophy to concerns about practical reason see Sullivan (1987).
2. See Toulmin's (1983, pp.398-400) distinction between 'topical' and 'analytical' arts and his extension of 'topical analysis' to science.
3. Note that topical analysis of scientific disputes presupposes that issues cannot be settled through appeals to objective, uniformly applicable standards that determine the certain empirical truth of the matter. Absence of purely objective truth standards, however, does not mean those disputes are fundamentally groundless. Instead, the grounds for argument and judgment turn out to be humanly created values, attitudes, and beliefs. Topical analysis examines how arguments incorporate those grounds to constrain judgments of comparative 'reasonableness' or 'appropriateness' within particular, practical situations.
4. Classical theory for issue, or 'stasis,' analysis was most fully developed for forensic pleading in courts of law, but rhetorical and dialectical treatises make it clear that the fundamental issues were thought more widely applicable and included political and philosophical disputations. I contended that the problems or ambiguities giving rise to argumentation in science also are structured according to those fundamental kinds of issue. For discussion of classical stasis theory and its adaptation for use in the analysis of scientific argumentation see Prelli (1989, chapters 4 and 8.)
5. For an introduction to theory of rhetorical topics and an inventory of *topoi* specific to deliberations about science see Prelli (1989, chapters 5 and 9).
6. *Topoi* are definite in structure but quite indefinite in circumstantial details. They become meaningful only through application to specific circumstances. Of course, any specific applications are always potentially arguable. For instance, one might question whether the technical value of 'precision' applies at all, or disagree with its proffered meaning, or adduce, say, explanatory power as a more important standard for adjudicating issues within the circumstances of the specific case.
7. Based partly on the DNA case study which appeared in Prelli, L.J., *Rhetoric of Science: Inventing Scientific Discourse* (1989), University of South Carolina Press.
8. Due to the brevity of the announcement, I shall refer to its paragraphs by number, 1-14.
9. Watson and Crick's *interpretive* framing of the structural problem or ambiguity was constrained by their chosen methodological orientation. That is typical in rhetoric about science. They wanted to solve the molecule's three dimensional structure through molecular model building, an approach Linus Pauling had earlier used to solve the structure of protein. As Lagerkvist (1998, pp.121-122) put it, Watson and Crick 'had learned from Pauling that by knowing the structure of the building blocks, the distance between atoms, and the angles between the bonds that held them together, one could build meaningful molecular models.' The more restrictions rigorously built into the model, the less likely it would turn out to be spurious. In contrast, the scientists

who wrote the two articles published along with Watson and Crick's piece had framed the molecule's structural ambiguities as an *evidential* problem and sought a solution through careful experimental work and painstaking mathematical analysis of results. For them, model building was secondary to securing reliable experimental data. For more discussion of how methodological orientations constrain scientists' discourse see Prelli (1989, pp.250-256; 1996).

10. Pauling and Corey published a one-paragraph announcement (1953a) in *Nature*, with a detailed elaboration (1953b) following shortly after in another forum.

11. Furburg had corrected Astbury's error of putting the sugars in parallel, rather than perpendicular, relationship to the bases - an error that Pauling and Corey had repeated in their reliance on Astbury's work.

12. How Watson and Crick came about these and other experimental data raised some controversy about scientific ethics. Clearly, they had benefited from Franklin's unpublished work. Wilkins had shown Watson one of Franklin's pictures of DNA in the B-form, which Watson later said confirmed his view that the structure was helical (Watson, 1968, pp.106-107; Judson, 1979,pp.159-161). Max Perutz also gave Watson and Crick a research report that Franklin had submitted to the Biophysics Committee of the Medical Research Council. That report contained useful information about the A-form, the transition of DNA from A-form to B-form, and related data. According to Crick (Judson, 1979, pp.164-67), the report helped him to see the molecule's dyadic symmetry. Watson (1969, 1539; 1968, pp.114-15) acknowledged its importance in solving the DNA structure.
Since Franklin had no prior knowledge of either communication, questions arose about ethics. On those issues see Lwoff (1968) and Chargaff (1968), as well as Perutz's (1969) account of his release of the report.

13. With simultaneous publication imminent, Wilkins recommended that Watson and Crick drop from their original draft of the announcement as 'a bit ironical' the sentence, 'It is known that there is much unpublished experimental material' (Judson, 1979, p.178).

14. Within a few weeks after publication of their announcement, Watson and Crick (1953b) published an article that elaborated the genetic implications of the model. They decided to present their claims more boldly after they realized how strongly the King's College evidence supported their structure. As Crick (1974, p.766) put it: 'The main reason was that when we sent the first draft of our initial paper to King's College we had not yet seen how strongly the X-ray evidence supported our structure.' As a consequence, 'We were ... delighted to see how well their evidence supported our idea. Thus emboldened, Watson was easily persuaded that we should write a second paper.'

15. Judson (1979, pp.184-93) surveys the important work that immediately followed publication of the double helix.

16. Several scientists were much less convinced with the second paper and other subsequent efforts to work out the details of replication. Max Delbruck pointed to these complex problems when he said of Watson and Crick's structure: 'In retrospect, what the denouement was, was that both the principle of replication and the principle of readout are very simple, and the actual machinery for doing it is immensely complex. That's the way it has turned out' (Quoted in Judson, 1979, pp.60-61). Chargaff said of the second paper, 'I was much less in agreement with the scheme for DNA replication proposed in the second note; and even now, twenty years later, I cannot say that I am reconciled with it completely, the mechanism of DNA synthesis *in vivo* still being obscure to me' (Chargaff, 1974, p.768).

6. REFERENCES

Bazerman, C. (1981 What Written Language Does: Three Examples of Academic Discourse. *Philosophy of the Social Sciences, (11)* 361-87.

Burnham, J. C. (1987). *How Superstition Won and Science Lost: Popularizing Science and Health in the United States.* New Brunswick, NJ: Rutgers University Press.

Campbell, J.A, & Bensen K.R. (1996). The Rhetorical Turn in Science Studies. *Quarterly Journal of Speech,* (82) 74-91.

Chargaff, E. (1968). A Quick Climb Up Mount Olympus (rev. of *The Double Helix*, by James D. Watson), *Science, (159)* 1448-1449.

Chargaff, E. (1974). Building the Tower of Babble. *Nature (Lond) (248)* 776-779.

Chargaff, E. (1978). *Heraclitean Fire: Sketches From a Life Before Nature.* New York: Rockefeller University Press.

Crick, F.H.C. (1974). The Double Helix: A Personal View. *Nature (Lond)*, (248) 766-771.

Crick, F.H.C. (1988). *What Mad Pursuit: A Personal View of Scientific Discovery.* New York: Basic Books.

Dewey, J (1929). *The Quest for Certainty: A Study of the Relation of Knowledge and Action.* NY: Minton, Balch, and Company.

Fisher, W.R. (1994). Narrative Rationality and the Logic of Scientific Discourse. *Argumentation*, (8) 21-32.

Franklin, R.E. & Gosling, R.G. (1953). Molecular Configuration in Sodium Thymonucleate. *Nature, (Lond)* (171) 740-741.

Gross, A.G. (1990). *The Rhetoric of Science.* Cambridge, MA: Harvard University Press.

Halloran, S.M. (1984). The Birth of Molecular Biology: An Essay in the Rhetorical Criticism of Scientific Discourse. *Rhetoric Review*, (3): 70-83.

Judson, H.F. (1979). *The Eighth Day of Creation: Makers of the Revolution in Biology.* NY: Simon and Schuster.

Kuhn, T.S. (1970). *The Structure of Scientific Revolutions.* Chicago: University of Chicago Press.

Lagerkvist, U. (1998). *DNA Pioneers and Their Legacy.* New Haven, CT: Yale University Press.

Lasch, C. (1995). *The Revolt of the Elites and the Betrayal of Democracy.* New York: London.

Lwoff, A. (1968). Truth, Truth, What is Truth (About How the Structure of DNA Was Discovered)? *Scientific American*, (219): 133-138.

Miller, C.R. (1992). Kairos and the Rhetoric of Science, in S Witte, N Nakadate, & R D. Cherry (eds.), *A Rhetoric of Doing.* Carbondale, IL: Southern Illinois Press.

Nelkin, D. (1987). *Selling Science: How the Press Covers Science and Technology.* New York: Freeman.

Pauling, L (1974). Molecular Basis of Biological Specificity. *Nature.(Lond.)* (248) 769-771.

Pauling, L, & Corey R.B.. (1953a). Structure of Nucleic Acids. *Nature.(Lond.)* (171) 346.

Pauling, L, & Corey R.B. (1953b). A Proposed Structure for the Nucleic Acids. *Proceedings of the National Academy of Sciences*, (39) 84-97.

Pera, M, & Shea, W.R. (eds.). (1991). *Persuading Science: The Art of Scientific Rhetoric.* Canton, MA: Science History Publications.

Perutz, M.F. (1969). Letter to the editor. *Science*, (164) 1537-1539.

Prelli, L.J. (1989). *A Rhetoric of Science: Inventing Scientific Discourse.* Columbia, SC: University of South Carolina Press.

Prelli, L.J. (1990). Rhetorical Logic and the Integration of Rhetoric and Science. *Communication Monographs, (57)* 315-22.

Prelli, L.J. (1996). Empirical Diversity, Interdependence, and the Problem of Rhetorical Invention and Judgment: The Case of Wife Abuse Facts. *Communication Theory, (4)* 406-429.

Shamos, M.H. (1995). *The Myth of Scientific Literacy.* New Brunswick, NJ: Rutgers University Press.

Stearney, L.M. (1996). Sex Control Technology and Reproductive 'Choice': The Conflation of Technical and Political Argument in the New Science of Human Reproduction. *Communication Theory, (6)* 388-405.

Sullivan, W.M. (1987). After Foundationalism: The Return to Practical Philosophy, in E. Simpson (ed.), *Anti-Foundationalism and Practical Reasoning: Conversations Between Hermeneutics and Analysis.* Edmonton, Alberta: Academic Printing, 21-44.

Toulmin, S. (1983). Logic and the Criticism of Arguments, in J.L. Golden, F.B. Goodwin, and W E, Coleman, (eds.), *The Rhetoric of Western Thought. (3rd ed.).* Dubuque, IA: Kendall-Hunt, 391-401.

Toulmin, S. (1988). The Recovery of Practical Philosophy. *American Scholar, (57)* 337-352.

United Kingdom. (1995). *Wolfendale Report, Section 1.8.*

United States of America. (August 3, 1994). *Science in the National Interest.*

Waddell, C. (1990). The Role of *Pathos* in the Decision-Making Process: A Study in the Rhetoric of Science Policy. *Quarterly Journal of Speech, (76)* 381-400.

Watson, J.D. (1968). *The Double Helix.* NY: New American Library.

Watson, J.D. (1969). Letter to the editor. *Science, (164)* 1539.

Watson, J.D. and. Crick F.H.C.. (1953a). A Structure for Deoxyribose Nucleic Acid. *Nature, 171:* 737-738.

Watson, J.D. and F.H.C. Crick. (1953b). Genetical Implications of the Structure of Deoxyribonucleic Acid. *Nature,(Lond.) (171)* 964-967.

Weaver, R.M. (1985). *The Ethics of Rhetoric*. Davis, CA: Hermagoras Press.

Wilkins, M.H.F., Stokes, A.R., and Wilson, H.R. (1953). Molecular Structure of Deoxypentose Nucleic Acid. *Nature,(Lond.) (171)* 738-740.

R. ECKERSLEY

5. POSTMODERN SCIENCE:
THE DECLINE OR LIBERATION OF SCIENCE?

1. INTRODUCTION

I sometimes think that the appeal of postmodernism to many people, myself included, is that it relieves us of the effort of trying to make sense of a world that no longer seems to make sense. This would have profound implications for science, which is, after all, about making sense of the world.

It is not that simple, of course. But postmodernity does pose interesting and important challenges for science – and for science communication and the public understanding of science. Science communication is, and must be, about much more than 'selling' science to the public. Fundamentally, it concerns the relationship between science and society, and it has a powerful role in shaping this relationship and also in what science is done and how it is used, not just economically, but culturally too. This means science communication is also closely associated with science policy issues.

In this chapter, I want to look broadly at several different - but, I think, related - aspects of how science could influence and be influenced by cultural developments, especially in modern Western societies. In particular, I want to look at the possible impact on science of the cultural changes associated with postmodernity, the relationship between science and material progress, and, finally, the potential for a reconciliation between scientific and spiritual world views. All have far-reaching implications for humanity. And depending on how these matters are played out, we could see, in the 21st century, either the decline or liberation of science.

2. POSTMODERNITY AND SCIENCE

Postmodernity (or late modernity, as some scholars prefer to call it) describes a world coming to terms with its limitations, including the recognition that the 'modern' dream of creating a perfect social order is ending, and that some of our problems may be insoluble. Postmodernity is marked by ambivalence, ambiguity, relativism, pluralism, fragmentation, contingency and paradox. There are no grand narratives or creeds that define who we are and what we believe, but a multiplicity of them.

Science and technology are among the key instruments of the modernist vision. As Anthony Elliott (1996) states:

> Science, bureaucracy and technological expertise serve in the modern era as an orientating framework for the cultural ordering of meaning. (pp. 18 - 19).

S.M. Stocklmayer et al. (eds.), Science Communication in Theory and Practice, 83–94.

This changes in a postmodern world.   Elliott argues that the vision of the Enlightenment has faded.

> The grand narratives that unified and structured Western science and philosophy... no longer appear convincing or even plausible

From a postmodern perspective, he says,

> knowledge is constructed, not discovered; it is contextual, not foundational.

Elliott (1996) argues that knowledge generated by experts and institutions is no longer equated with increasing mastery and control of the social order.   In fact, he says, the advance of modernisation is increasingly equated with the production of risks, hazards and insecurities on an unprecedented global scale.

> Put more accurately, technological knowledge and control of the social world today are as much about managing socially produced risks and dangers which are worldwide in their consequences as about unbounded mastery in the service of political domination. (pp. 66-70)

The profound paradox of our situation is well described by Marshall Berman (cited by Elliott, 1996), who said:

> To be modern is to find ourselves in an environment that promises us adventure, power, joy, growth, transformation of ourselves and the world - and, at the same time, that threatens to destroy everything we have, everything we know, everything we are. (p. 11)

So we can see that there are two aspects to the postmodernist critique of science: epistemic relativism; and science as a two-edged sword.

Scientists are most hostile to the first charge - that scientific knowledge is culturally adulterated.   I do not entirely agree with this assertion.   Scientific knowledge does transcend its cultural context; science does 'advance' in a way that is, I think, unique.   But scientific knowledge is never the whole truth or an absolute, immutable truth.   And what science is done, and how its results are applied, are powerfully determined by its cultural context.

So, given that we choose into which corner of the dark cavern of the unknown we shine the light of scientific inquiry, and given that we will never light up everything, then we do need to acknowledge the degree to which what we see depends on what influences our choice of where to look and what to look for - that is, on who we are and what we believe.   This degree of cultural construction depends on the science: smaller in the case of the physical sciences, larger in the social; lesser in pure science than in applied.

The second charge against science - that it is a mixed blessing - is uncontestable, and doesn't need elaboration.   This applies to specific products of science (technologies) such as nuclear energy, pesticides or genetic modification, or more broadly to the whole relationship between science and material progress - a subject to which I will return later.

There is a second factor which could compound the effect of postmodern thinking on science: the possibility that science may have to confront its own intrinsic limitations.

John Horgan (1996) has argued that we must accept the possibility that the great era of scientific discovery is already over. He is not referring to applied science, which still has an abundance of problems to solve, but what he calls 'science at its purest and grandest, the primordial human quest to understand the universe and our place in it'. Horgan develops an idea propounded by Gunther Stent in *The Coming of the Golden Age: A View of the End of Progress*, published thirty years ago. Stent argued that if there are any limits to science, any barriers to further progress, then science may well be moving at unprecedented speed just before it crashes into them. When science seems most muscular, triumphant, potent, that may be when it is nearest death, Stent said.

> Indeed, the dizzy rate at which progress is now proceeding makes it seem very likely that progress must come to a stop soon, perhaps in our lifetime, perhaps in a generation or two.

Horgan implies three different reasons for this view. One reason is that all the major discoveries–or should we call them 'constructions'?–may have been already made:

> Now that science has given us its Darwin, its Einstein, its Watson and Crick, the prospect arises that further research will yield no more great revelations or revolutions but only incremental, diminishing returns.

(He discusses, but dismisses, the claim that scientists thought this about physics last century.) Another reason is that even seemingly open-ended sciences like physics inevitably confront physical, financial and even cognitive limits: modern physics, for example, is becoming increasingly difficult for anyone, even physicists, to comprehend. A third factor is the intrinsically indeterministic nature of many natural phenomena - that is, they are unpredictable and apparently random–making them resistant to scientific analysis. The work emerging from chaos and complexity theories demonstrates that science, when pushed too far, culminates in incoherence.

I am not necessarily endorsing Horgan's arguments, only suggesting they deserve consideration in looking at the future of science.

3. POSTMODERN SCIENCE

So science is being assailed by two forces: the first, postmodernism and its challenge to science's social and intellectual authority; the second, science's own 'limits to growth'. What will be the consequences?

While technological innovation will continue apace, science will cease to be the defining and dominant feature of our society. It will co-exist, often uncomfortably, with irrationalism, superstition and other belief and knowledge systems. In losing its ideological dominance as the source of progress, science is losing its own internal coherence, and the philosophy and culture that have held it together. While good science will remain rigorous and empirical, this will be more a question of professional ethics and sheer pragmatism - this science delivers the best results - than the sort of ideal represented by sociologist Robert Merton's four norms of science: universalism, communism, disinterestedness and organised scepticism.

Like everything else, science is fragmenting. Much more openly and unequivocally than in the past, science today serves different masters and different purposes. Its culture and norms become those of its users. Thus, it is increasingly meaningless to talk about a single form of scientific progress, or about attitudes to science in any generic sense. Public opinion about science depends on which public and which science. The epigraph on the United States National Academy of Sciences building in Washington–

> To science, pilot of industry, conqueror of disease, multiplier of harvest, explorer of the universe, revealer of nature's laws, eternal guide to truth

will, with its implied congruence and attainability of all these goals, its unified vision of progress, become a quaint anachronism in the postmodern world.

This is already apparent from surveys of how people perceive science and technology. They are ambivalent and contradictory in their views - and also discerning. Take, for example, a study I initiated several years ago, under the auspices of the Australian Science, Technology and Engineering Council, which explored young people's hopes and fears for Australia in the year 2010: a key finding was the extent to which views on science and technology were embedded in a wider social context (Eckersley, 1999). The role young people saw for science and technology changed markedly between their expected and preferred futures.

Young people are not so much *against* science and technology. Indeed, they acknowledge their importance in achieving a preferred future. But they are astute enough to realise science and technology are tools, and their impacts depend on who controls them and whose interests they serve. They *expected* to see new technologies used further to entrench and concentrate wealth, power and privilege. They *wanted* to see new technologies used to help create closer-knit communities of people living a sustainable lifestyle.

For example, young Australians (aged 15-24) were asked in one poll question to agree or disagree with nine specific statements about science and technology. The responses showed that:

> Young people believed science and technology offered the best hope for meeting the challenges ahead (69%), but also that they were alienating and isolating people from each other and nature (53%).
> They believed that computers and robots were taking over jobs and increasing unemployment (58%), and a significant minority believed that they would eventually take over the world (35%).
> They were more likely to think that governments would use new technologies to watch and regulate people more (78%) than they were that new technologies would strengthen democracy and empower people (43%).
> They expected science to conquer new diseases (87%), but not that it would find ways to feed the growing world population (39%), or solve environmental problems without the need to change lifestyles (45%).

In another question, young people were asked to nominate which of two positive scenarios for Australia in 2010 came closer to the type of society they both expected and preferred:

> A fast-paced, internationally competitive society, with the emphasis on the individual, wealth generation and 'enjoying the good life'. Power has shifted to international organisations and business corporations. Technologically advanced, with the focus on economic growth and efficiency and the development of new consumer products.

> A 'greener', more stable society, where the emphasis is on cooperation, community and family, more equal distribution of wealth and greater economic self-sufficiency. An international outlook, but strong national and local orientation and control. Technologically advanced, with the focus on building communities living in harmony with the environment, including greater use of alternative and renewable resources.

Almost two thirds (63%) said they expected the first, 'growth' scenario. However eight in ten (81%) said they would prefer the second, 'green' scenario. About a third (35%) expected the 'green' scenario, and 16% preferred the 'growth' scenario.

One possible consequence of postmodernity is that science will become a greatly diminished cultural influence in our lives and in national affairs (even while we continue to embrace its products). For example, Horgan (1996) sees the limitations of science contributing to a growing reluctance by the public to support science, and even to the rise of anti-scientific sentiments. He notes that Oswald Spengler foresaw the disillusionment with science in *The Decline of the West*, published in 1918: Spengler predicted that the demise of science and the resurgence of irrationality would begin at the end of the millennium. As scientists became more arrogant and less tolerant of other belief systems, notably religions, he believed society would rebel against science and embrace religious fundamentalism and other irrational systems of belief.

There are signs that this might be happening, although public sentiment has not so much swung against science and technology as shifted towards superstition and fundamentalism. For example, Americans view science and technology as the engines of the past century's economic prosperity and the main reasons for the improvements in their well-being, and are optimistic about further gains in the next century. Yet they also express misgivings about the way their country has changed culturally and spiritually (Pew, 1999). Asked in a recent poll what was more important, encouraging a belief in God or encouraging a modern scientific outlook: 78% of Americans chose 'a belief in God', and only 15% 'a modern scientific outlook' (Washington Post/Kaiser/Harvard, 1998). Over a third (36%) believe the Bible is the actual word of God, to be taken literally word for word, while almost half (48%) believe it is the inspired word of God, but not everything in it should be taken literally. Only 14% regarded the Bible as an ancient book of fables, legends, history and moral precepts recorded by man.

But there are also other possibilities. In the early 1990s, I wrote in essays for the Australian Commission for the Future and *The Futurist* that science could play a crucial role in achieving the sort of cultural or values shift necessary to address 21st century challenges (Eckersley 1992, 1993). But in effecting change, science must itself be changed. While remaining rigorous, science must become intellectually less arrogant, culturally better integrated and politically more influential. Science must become more tolerant of other forms of reality, other ways of seeing the world. It must become less remote from public culture, with a steadier and readier flow of influence between the two - in both directions. And it must contribute more to setting political agendas. Science communication has a pivotal role in these changes.

I did not realise then how postmodern this perspective was. It represents perhaps the best outcome for postmodern science. There are also signs that this is happening! It is from this perspective that postmodernism can liberate science. By forcing us to acknowledge that science is not a dispassionate, value-free search for objective knowledge about nature and society, that it is imbued with the subjective and conditioned by its social and cultural environment, science becomes more pluralistic and flexible.

Science can shed its close association with 'progress' as we currently define it, and openly associate itself with other social goals. Science and scientists have, after all, been the driving force behind the modern environmental movement. We could see the growth of a 'transformational science', a highly interdisciplinary style of research that would draw its inspiration, its coherence, from a shared ideal to use science to achieve a transition from a society defined by economic growth and a rising material standard of living, to one that offers a high, equitable and sustainable quality of life. While the research would continue to be directed towards practical outcomes, it would be defined and guided by this transformational vision of sustainable development.

4. SCIENCE, INNOVATION AND PROGRESS

The concept of sustainable development poses the most formidable challenge so far to that of material progress, the defining characteristic of Western society for the past few centuries. Within the paradigm of material progress, the role of science has increasingly come to be seen as the engine of progress, the source of innovation that drives economic growth and wealth creation.

A few decades ago, innovation was seen as a linear process, a pipeline: research led to invention which was commercialised to become innovation. Since then, this model has been replaced by a succession of more complex models that emphasise the dynamic and social nature of innovation, which involves researchers, producers, marketers and customers in an iterative, interactive process (Bessant and Dodgson, 1996). The contributions of research to this process go beyond being a source of invention and are often indirect and diffuse.

Innovation theories seem to be continuing to evolve in the light of new experience. A recent book, *The Innovator's Dilemma*, focuses on 'disruptive technologies' and argues that it is only by ignoring current customers and disobeying seemingly sound management practices that drastic innovation can be harnessed (Christensen, 1999). Not only are the market applications for disruptive technologies unknown at the time of their development, they are unknowable, and, by and large, they are initially embraced by the least profitable customers in a market.

While innovation theory has come a long way in recent decades, our notions of progress seem still to be based on a model similar to the old linear model of innovation (Eckersley, 1998). Progress is like a pipeline: pump more wealth in one end, and more welfare will flow out the other. Current government policy is underpinned by the belief that wealth creation comes first: economic growth increases our capacity to meet environmental and social objectives, as well as raising material standards of living. This notion provides the framework of debate, which then centres around how the fruits of growth should be distributed.

The fundamental assumptions about economic growth - that it enhances welfare or well-being and that it is environmentally sustainable - are rarely highlighted or explored. However, the belief in the primacy of wealth creation is wrong: if the processes by which we pursue economic growth do more damage to the social fabric and the state of the environment than we can repair with the extra money, then we are still going backwards (even assuming we can fully identify, cost and repair the impacts). 'Efficiency' in generating wealth may well mean 'inefficiency' in improving overall quality of life.

The rationale for continuing economic growth in rich countries seems flawed in several important respects:

> it overestimates the extent to which past improvements in well-being are attributable to growth; it reflects too narrow a view of human well-being, and fails to explain why, after 50 years of rapid growth, so many people today appear to believe life is getting worse; and it underestimates the gulf between the magnitude of the environmental challenges we face and the scale of our responses.

The issue of contention in the debate about progress is not simply growth versus no-growth. That growth is better than recession in generating jobs - the main political justification for promoting growth - is insufficient reason for not looking much more closely at *what* is growing, what *other effects* this growth is having, and what *alternatives* might exist. We need to examine more critically the whole basis on which progress is currently defined, measured and achieved. Instead of just 'going for growth', and focusing science on that end, we need to be 'going beyond growth'. To suggest this is not to be 'anti' the economy, business or scientific research and technological development; it is to argue that these activities need to be driven by different values towards different ends.

5. SCIENCE AND RELIGION

I touched earlier on the tension between science and religion. Given religion's importance to values, let me come back to this issue. The relationship between science and religion today hangs in balance between conflict and concurrence. The outcome will depend on our interpretations of what both are.

The religious experience is not easy to articulate. Theologians argue that God is beyond images and beyond thought. Thomas Aquinas said that we know God best when we come to the point of knowing that we don't know him. A Sanskrit text, the Upanishad, says of Brahman (the ultimate reality, or Self, from which the world was created): 'Brahman is unknown to those who know it and is known to those who do not know it at all'. The novelist, Morris West, a devout Catholic, once said:

> I don't know who or what God is but I do know that there is a relationship between me and the Cosmos and its origins. I'm part of it.

Charles Birch, a biologist and theologian, also emphasises the 'relational' nature of God (Birch, 1993, p.236). God, he says, 'is internally related to all that is'. 'God is to the world as self is to the body.' As I understand this, he is saying our relationship to God is personal, but it is an internal relationship, not a relationship to something or someone else; there is no 'other'.

I define spirituality as a deeply intuitive sense of relatedness or connectedness to the world and the universe in which we live. Religions are social institutions built up around a particular spiritual metaphor, or set of metaphors, for this relationship. Religions may be socially necessary and desirable to obtain the greatest social and personal benefits from a sense of the spiritual - meaning, fulfilment, virtue. However, they can be made so rigid and sclerotic by institutional inertia, and by layers of bureaucracy, politics and corruption, that their spiritual core withers. When this happens, they become self-serving institutions lacking any higher purpose; worse, they can become potent ideologies of oppression and abuse.

Science also uses metaphors to describe the world. These days, cosmology is full of terms like black holes, worm holes, quantum foam. We are learning that science and religion use different metaphors to describe the same world, or different dimensions of the same world. (Some metaphors, such as Gaia, the notion of the Earth as a single, self-regulating living system or organism, can even be both scientific and religious).

Here are two scientific descriptions of the world. The first comes from biologist Richard Dawkins (1995):

> In a universe of electrons and selfish genes, blind physical forces and genetic replication, some people are going to get hurt, other people are going to get lucky, and you won't find any rhyme or reason in it, nor any justice. The universe that we observe has precisely the properties we should expect if there is, at bottom, no design, no purpose, no evil and no good, nothing but pitiless indifference.

The second is from physicist Paul Davies (1995):

> The true miracle of nature is to be found in the ingenious and unswerving lawfulness of the cosmos, a lawfulness that permits complex order to emerge from chaos, life to emerge from inanimate matter, and consciousness to emerge from life....(T)he universe (is) a coherent, rational, elegant and harmonious expression of a deep and purposeful meaning.

The two views represent the extremes of the modern scientific worldview. According to the first, we are doing what all species have ever done: to do as well as possible, to sequester for ourselves as much of the earth's resources as we possibly can. According to the second, we are part of an awesome evolutionary pattern that has seen, in the space of some 12 billion years, the emergence of a universe that can wonder and marvel at itself. I do not think the two perspectives are irreconcilable, and simply reflect different dimensions of the evolution of life - Dawkins focusing on living organisms and their struggle for survival, Davies on a more abstract cosmology.

Western culture has been deeply influenced by the old, Newtonian model of a dead, mechanical, clockwork universe. It has yet to absorb the significance of the new model, one of a dynamic cosmic network of forces and fields, of an 'undivided, flowing wholeness' - to use physicist David Bohm's words - that is far more compatible with a spiritual sense of connectedness to the universe.

The Nobel laureate, Steven Weinberg (1994), has argued that life as we know it would be impossible if any one of several physical quantities had slightly different values. For example, the vacuum energy or cosmological constant appears to need to be fine-tuned to an accuracy of about 120 decimal places for life to exist in the universe. Weinberg acknowledges that opinions differ on the degree of this fine-tuning. He also says this does not necessarily mean that 'life or consciousness plays any special role in the fundamental laws of nature'. Still, it raises an intriguing question: is this the razor's edge of probability, or exquisite precision engineering?

So spirituality is the intuitive sense of what science seeks to explain rationally. For me, the significance of all this is not that there is some Divine Purpose or Supreme Being somewhere 'out there' that gives meaning to life. Rather this understanding, or awareness, of our relationship with the Cosmos fosters a sense of deeper purpose, or meaning, within ourselves.

The anthropologist, Clifford Geertz, said that:

> Whatever else religion does, it relates a view of the ultimate nature of reality to a set of ideas of how man is well-advised...to live. (cited in Novak, 1994, p.111).

It has often been said that science, while also offering a view of 'the ultimate nature of reality' lacks the moral dimension. Yet research in a wide range of disciplines - from psychology and physiology, epidemiology and sociology, to ecology and cosmology - does provide guidance on how we ought to live, guidance of a kind that is compatible and consistent with religious teaching. But in both realms - science and spirituality - we are operating at the very limits of our capacity to comprehend 'the grand scheme of things'. At this conceptual level, our view is highly subjective, we can only express ourselves in metaphors; the moral lessons can only be human interpretations, not laws of science or of God.

Religion faces a growing tension that will bear mightily on its future: a tension between developing new, or renewed, 'transformational' religions and retreating to old, fundamentalist faiths. The former would use metaphysical metaphors and practices attuned to our times and our modern, scientific understanding of the world; the latter offer rock-solid certainties in a time when these can be enormously destructive. I don't mean, in talking about this tension, to sideline current mainstream faiths, but rather to suggest they will be caught up in it, and could be profoundly shaped by it. The danger with fundamentalism is that it mistakes the religious 'metaphor' for the spiritual 'truth', and so cedes too much power to those who claim to speak on God's behalf. On the other hand, more 'modern' concepts of God, while philosophically compelling, may be too abstract to meet the human yearning for spiritual comfort and moral authority. Still, this path seems to me to offer the best prospects of a better future - harder, undoubtedly, but more likely in the long run to lead to a peaceful, equitable and sustainable world. Science has a role in encouraging us to take this path.

The new religions would transcend, rather than confront, the powerful individualising and fragmenting forces of postmodernity. One of the most exciting ideas to emerge from recent postmodern scholarship is that we have the opportunity, however small, of becoming truly moral beings, perhaps for the first time in history (Bauman, 1995, pp.10-43, pp.256-288). That is, we have, each of us, the opportunity to exercise genuine moral choice and to take responsibility for the consequences of those choices, rather than accepting moral edicts based on some grand, universal creed and handed down from on high by its apostles. Bauman (1995) writes:

> The denizens of the postmodern era are, so to speak, forced to stand face-to-face with their moral autonomy, and so also with their moral responsibility. This is the cause of moral agony. This is also the chance the moral selves never confronted before. (p. 43)

This seems close to what theologians call the doctrine of 'primacy of conscience'. It is an immense challenge, and it may well be asking too much of us. But the ideal is there, if often hidden, in both religious teaching and science.

Human well-being is associated with the deep and enduring personal, social and spiritual attachments that give our lives a moral texture and a sense of meaning - of self-worth, belonging, identity, purpose and hope. Psychologists have shown that positive life meaning is related to strong religious beliefs, self-transcendent values, membership of groups, dedication to a cause and clear life goals. Bruce Headey and Alex Wearing (1992, p. 191) say in their book, *Understanding Happiness*, that: 'A sense of meaning and purpose is the single attitude most strongly associated with life satisfaction'. The psychologist, Martin Seligman (1990), argues that one necessary condition for meaning is the attachment to something larger than the self, and the larger that entity, the more meaning people can derive. 'The self, to put it another way, is a very poor site for meaning.'

Meaning in life need not be religious. Many people today find it in the pursuit of personal goals – in careers, sport or family, for example. But spirituality offers something deeper. It is central to the age-old questions about the meaning of life: Who am I? Where have I come from? Why am I here? It represents the broadest and deepest form of connectedness. It is the most subtle, and so easily corrupted by societies, yet perhaps the most powerful. It is the only form that transcends our personal circumstances, social situation and the material world, and so can sustain us through the trouble and strife of mortal existence.

Morris Berman (1990) concludes his book, *Coming to Our Senses: Body and Spirit in the Hidden History of the West*, with these words:

> Something obvious keeps eluding our civilisation, something that involves a reciprocal relationship between nature and psyche, and that we are going to have to grasp if we are to survive as a species. But it hasn't come together yet, and as a result, to use the traditional labels, it is still unclear whether we are entering a new Dark Age or a new Renaissance. (p. 344)

How we think of science and its relationship with society, the economy and culture, including religion, will have a large bearing on whether we grasp this 'something' and so which way we go. And, as I said in the Introduction, this question is, fundamentally, a challenge for science communication.

6. CONCLUSION

In *Biology and the Riddle of Life*, Charles Birch (1999) says science inevitably leads to mechanical analyses. Is there nothing more to be said, he asks:

> I think there is. It is to propose that there are two points of view – *the inside and the outside, the subjective and the objective, from within and from without*.....There is an enormous gap between what science describes and what we experience.....(T)he solution to the riddle of life is only possible through the proper connection of the outer with the inner experience. (p. 58)

I have argued that postmodernity could see the decline of science, or its liberation. By liberation, I mean abandoning, or at least qualifying, our notions of scientific knowledge as objective and value-free - as discovered, rather than constructed. And I also mean breaking free of a narrow, limited view of science, particularly by governments, that sees the prime objective of science policy as harnessing scientific research and education ever-more closely to the task of growing the economy - to the pursuit, in other words, of material progress.

In taking a broader view of the relationship between science and society - and so of science communication - I also discussed the role of science in achieving a high, equitable and sustainable quality of life, rather than a life that is merely materially richer, and in strengthening a sense of the spiritual.

The current dominant view of science as an economic tool is such an impoverished vision. Science has much, much more to offer humanity than that.

National Centre for Epidemiology and Population Health at the Australian National University, Canberra, Australia.

7. REFERENCES

Bauman, Z. (1995). *Life in Fragments: Essays in Postmodern Morality*. Oxford, UK: Blackwell.
Berman, M. (1990). *Coming to Our Sense: Body and Spirit in the Hidden History of the West*. London: Unwin Paperbacks.
Bessant, J. and Dodgson, M. (1996). *Effective Innovation Policy: A New Approach*. New York: Routledge.

Birch, C. (1993). *Regaining Compassion for Humanity and Nature*. Sydney: University of New South Wales Press.

Birch, C. (1999). *Biology and the Riddle of Life*. Sydney: University of New South Wales Press.

Christensen, C. (1999). *The Innovator's Dilemma*. Boston: Harvard Business School Press.

Davies, P. (1995). Physics and the Mind of God: The Templeton Prize Address. *First Things (August/September)*, 31-35.

Dawkins, R. (1995). God's Utility Function. *Scientific American (November)*: 62-67.

Eckersley, R. (1999). Dreams and Expectations: Young People's Expected and Preferred Futures and Their Significance for Education. *Futures, 31*, 73-90.

Eckersley, R. (1998). Perspectives on Progress: Economic Growth, Quality of Life and Ecological Sustainability. In R. Eckersley (Ed). *Measuring Progress: Is Life Getting Better?* Collingwood, Victoria: CSIRO Publishing.

Eckersley, R. (1993). The West's Deepening Cultural Crisis. *The Futurist (November-December)*, 8-12.

Eckersley, R. (1992). Youth and the Challenge to Change: Bringing Youth, Science and Society Together in the New Millennium. *Apocalypse? No! essay series no. 1*, Melbourne: Australian Commission for the Future,

Elliott, A. (1996). *Subject to Ourselves: Social Theory, Psychoanalysis and Postmodernity*. Cambridge, UK: Polity Press.

Headey, B. & Wearing, A. (1992). *Understanding Happiness: A Theory of Subjective Well-being*. Melbourne: Longman Cheshire.

Horgan, J. (1996). The Twilight of Science. *Technology Review (July)*, 50-60.

Novak, P. (1996). *The World's Wisdom: Sacred Texts of the World's Religions*. Edison, N.J.: Castle Books. (First published 1994).

Pew Research Center. (1999). 'Public Perspectives on the American Century: Technology Triumphs, Morality Falters' and 'Americans Look to the 21st Century: Optimism Reigns, Technology Plays Key Role'. *1999 Millennium Survey 1&2*. www.people-press.org .

Seligman, M. (1990). Why is There So Much Depression Today? The Waxing of the Individual and the Waning of the Commons. In R.E. Ingram. (Ed). *Contemporary Psychological Approaches to Depression – Theory, Research and Treatment*. New York: Plenum Press.

The Washington Post/Kaiser/Harvard Survey Project. (1998). *American Values: 1998 Survey of Americans on Values*. Kaiser Family Foundation, www.kff.org,

Weinberg, S. (1994). Life in the Universe. *Scientific American (October)*, 22-27.

SECTION 2

THE EXPERIMENTAL APPROACH TO SCIENCE COMMUNICATION

D. SLESS AND R. SHRENSKY

6. CONVERSATIONS IN A LANDSCAPE OF SCIENCE AND MAGIC: THINKING ABOUT SCIENCE COMMUNICATION

1. INTRODUCTION

Communication is one of the contemporary 'success' words. It conjures up ideas of successful action and change: 'getting our message across', 'running public information campaigns', 'changing attitudes', 'raising awareness', 'setting the agenda', 'managing a public image'. These are the types of positive, success-orientated ways that our contemporaries talk about communication in public life.

Not surprisingly then, something called Science Communication has arisen wherever scientists, science journalists, public relations consultants, and science policy workers perceive a gap between their sense of science and its importance, on the one hand, and public perception of science, on the other hand. If there is a falling off of recruits to science courses, the answer is Science Communication. If people think scientists are nerds, mad, evil, or just boring, the answer is Science Communication. If the government cuts funding in scientific research, the answer is... and so on. In other words, science communication is seen as a tool for getting the job done. The more of it, the better!

Implicit in this view of science communication is a simple belief: communication works. Communication, it is believed, has the capacity to change people's attitudes and behaviour in any desired direction.

This simple belief - that communication does things to people - is so widespread, so widely accepted, that it seems beyond doubt and in no way problematic; indeed, it is 'common sense'. All that is needed is the skill and the money to go out and do it. Not surprisingly then, many have taken on the mission of science communication with great enthusiasm, conviction, and a strong desire to make it work. And success seems to be evident everywhere. Science centres and museums, column inches in the press, programs on TV - all these are evidence that science communication is out there and doing a good job. This is intuitively obvious - or so it seems.

We take a different view. We think that science communication activities in our society are like rainmaking ceremonies in nonscientific societies. We take this view for four reasons. First, from a scientific perspective, the evidence for the effectiveness of communication is about as strong as the evidence linking rainmaking ceremonies to the occurrence of rain. Second, from a rhetorical point of view, the rationale and arguments about the effectiveness of communication bear a striking similarity to the types of rationale and arguments that are used to support the efficacy of rainmaking ceremonies. Third, from an anthropological point of view,

S.M. Stocklmayer et al. (eds.), Science Communication in Theory and Practice, 97–105.

rainmaking ceremonies provide us with one of the best ways of understanding science communication practice. Fourth, from a philosophical point of view, to think about either rainmaking or communication in terms of cause and effect - ceremony performed, causing rain; message sent, causing public reaction - is not a productive way of thinking about either.

There are other ways of thinking about communication that open up some exciting possibilities for the future of science communication that do not involve a belief in a magical connection between communication and behaviour. This chapter, then, offers some thoughts not only on what science communication is not, but also on the more practical question of what science communication might be.

2. THE EVIDENCE

First, a brief account of mass communication research findings over the last fifty years. We have chosen this particular area of communication research because it most clearly illustrates the disjuncture between our common sense views of communication and the research evidence. The references we cite will lead you directly or indirectly to most of the major published sources in this field.

Our brief account begins with the rush of enthusiastic research into the effectiveness of communication which took place in the immediate years after the Second World War. One of the 'weapons' that distinguished this war from its predecessors was the use of mass media propaganda by all sides in the conflict - radio, newspapers, posters and leaflets and so on. The effect this public bombardment had on people was not quantifiable during the conflict, but afterwards researchers systematically and rigorously investigated the effectiveness of this kind of mass media. The assumption behind this research was that propaganda and other mass media products had a direct and powerful effect on attitudes, knowledge and behaviour.

By the late 1950s there were enough separate studies for the research community to stand back and look at the overall results, and make some generalisations. The results were disappointing, to say the least, for the believers in strong direct media effects. In a major review of the cumulative research into mass media effects up to 1960, Joseph Klapper showed that the search for major and predictable mass media effects had failed to find any such results (Klapper, 1960). Klapper's conclusion was that effects, if there were any, were small and unpredictable. It was as if someone in a pre-scientific society had gone back over all the rainmaking ceremonies of the last fifteen years and shown that only on very few occasions had it rained after a ceremony and, moreover, on most of those few occasions it had only drizzled for a minute or two.

Given the prevailing commonsense view of communication, most communication researchers were sceptical about Klapper's conclusions. They believed strongly that the media must have powerful effects of some kind; the problem was to develop a methodology that would find them. They argued that the early research had been flawed and simplistic.

Some suggested that the media effects might not be direct - like a hypodermic syringe - but indirect, through a chain of interpersonal connections where one person (an 'opinion leader') responded to a message and influenced the opinions of many others as a consequence. This was known as the two-step flow theory of media effects. Others argued that there was a complex hierarchy of effects, like snooker balls being set in motion by the first ball in the row; first a message sender should grab people's attention, then give them information; this changes their knowledge, which changes their attitudes, which in turn changes their behaviour.

Some researchers suggested that the effects were more subtle; the media didn't tell us what to think, but they told us what to think about; this was called the agenda-setting hypothesis. Finally, and running out of steam, some researchers suggested that the 'real' effects of the media were ideological. They induced within us our unconscious unreflected views of society, making us accept the dominant status quo as 'natural'.

But whatever theoretical standpoint was used to inform the research, the results kept on (and keep on) being similar to those early findings by Klapper. Weak or no correlations between media messages and public behaviour; unpredictable results; little or no changes in behaviour due to public information campaigns; and so on. Those who followed Klapper and tried to synthesise the findings from disparate studies came to the same types of conclusions. Whether the field was agenda setting (Blood, 1989), public information campaigns (McGuire, 1986), or advertising (Schudson, 1984), the results were remarkably similar in their overall pattern. Sometimes effects were observed, but they were usually small, and never predictable. There were occasional spectacular large apparent 'effects', like the panic that reportedly ran across the east coast of the USA when listeners to Orson Welles' dramatisation of H.G. Wells' *War of the Worlds* mistook the documentary realism of the drama for actual news. But these were extremely rare and unpredictable: the stuff ad men dream of but seldom, if ever, make happen for their clients.

Nonetheless, researchers still believed that there were powerful effects, if only they could be discovered. Denis McQuail, one of the most distinguished teachers in the field, commented in 1987:

> The entire study of mass communication is based on the premise that there are effects from the media, yet it seems to be the issue on which there is least certainty and least agreement. (p.251)

After half a century's research, the effects of communication remained as elusively out of reach as they had always been. Communication, like ceremonial rainmaking, has a poor track record in achieving desired outcomes. But none of this has had the slightest effect on practice. Most communication practitioners do not know about these findings, and those that do don't believe them. Those who conduct rainmaking ceremonies continue to perform them, even though it doesn't rain, and keeps on not raining.

3. THE RHETORIC

To understand why people continue to act in ways that are inconsistent with the evidence, it is worth looking at some of the arguments they offer to justify continuing the practice, and then the arguments they offer when, as inevitably happens, the practice does not yield the desired outcome. When you do this, the similarities between communication activity and rainmaking ceremonies become more apparent. What also becomes apparent is just how plausible the arguments are, sufficient in most cases to protect the activity from criticism.

Argument 1. People spend a lot of time and resources on it. When confronted by the negative evidence, the first argument that is offered in defence of communication practice is that it must be effective, otherwise why would big companies and governments continue to spend so much money on it?

Argument 2. I can cite examples where it has worked. This argument, by example, suggests that the practice must work because we all remember certain 'successes' in the field—a particular documentary or advertisement that had a huge impact.

Argument 3. The timing was wrong. When communication activity fails, there are many ways of plausibly explaining the failure. A popular one is that the program was put on at the wrong time: 'There were just too many other things going on; we couldn't compete with the competition for time and exposure'; 'We should have put our information program on at peak viewing time'.

Argument 4. Not enough resources. This popular argument comes in many guises. 'We just didn't have a big enough budget'. 'How can we compete with big business with all its resources when it comes to gaining public attention?'

Argument 5. The execution was flawed. One of the inevitable scapegoats for failure is the practitioner. 'The agency did a bad job'; 'The designers were incompetent'; 'The writing was awful'. Interestingly, this argument has practical consequences for what is called the communication 'industry': there is a high turnover of consultants. It is a common strategy for an organisation to sack an 'unsuccessful' communication practitioner or agency and employ a new one. In this way, current practice continues unchanged.

Argument 6. We did our best. When there is no-one to blame, no plausible explanation for failure, there is still the argument that 'we did what we could with what we had', and 'we will try again in the future'.

For all six arguments, there are parallels in rainmaking activities. Confront a rain maker with meteorological evidence which goes against his belief, and the answer would be an incredulous: 'If it didn't work, why would tribal elders invest in such large and powerful ceremonies? And anyway, we all remember the time when it started to rain just as the ceremony concluded. Of course it works! Everyone remembers that ceremony and the effect it had. It rained for a week. Sometimes the gods are not pleased, certainly, and refuse to give rain; or the omens are wrong. And, sadly, sometimes the witchdoctor does not have the right spirit - we soon get rid of him!'

All the above arguments defend the beliefs of traditional and conventional communication practice. The next and final argument adds a new dimension: it legitimates action.

Argument 7. We did it! Just as the fact of performing a rainmaking ceremony is taken as evidence of the process at work, so the very act of doing and making - producing the annual report, the web site, the exhibit or whatever - is considered sufficient evidence of the process of communication at work. Indeed, many practitioners would see the completion of the message as the completion of the work. Exhibitors get paid to create an exhibition, writers get paid to write, and so on. The visitor's experience or the reader's understanding is either taken for granted or part of somebody else's responsibility.

4. THE ANTHROPOLOGY

It is in this final argument and the visible evidence of practice that we can begin to understand what is happening from an anthropological point of view. Robert K Merton, a sociologist, (Merton, 1957) makes a useful distinction between what he calls 'latent' and 'manifest' functions. He argues that if we are trying to understand what is happening in a society by looking at the way in which people describe what they are doing, we may miss something important. If we look at a rainmaking ceremony too literally, we miss the fact that these ceremonies are major social rituals which, in a highly uncertain world, bring everyone together around a common purpose and set of beliefs. In other words, the work that rainmaking ceremonies do is not so much about making rain (the manifest function) as about maintaining social cohesion through shared ideas and experiences in an otherwise uncertain world full of mystery (the latent function).

We can better understand much of science communication when we think of it in this way. For example, science museums and centres are places which are manifestly set up to explain and teach non-scientists about science. But it is more useful to see them as places where scientific ideas and values are displayed and experienced. They are part of our society's ritual display, places where we tell each other what we regard to be important and enjoyable, and provide experiences that help establish shared rituals in areas where there is ignorance and mystery.

There are many places in our society where rituals take place. And each place has its own rules and conventions about how we take part in the rituals. Observe an adult and child walking through a science centre, a department store, and an art gallery. Different types of behaviour are acceptable in each location. The adult continually reminds the child how to behave in each context, how to partake in the ritual. To take part in the ritual the child has to learn certain conventions. In the science centre you can not only touch the exhibits, you can push buttons and pull handles, whilst making loud excited noises. In the department store, you can only touch things that you are going to buy, and you have to speak conversationally, and not run about and shout. In the art gallery you cannot touch anything, and you must talk very quietly, if at all.

Because we live in a largely secular society, we tend to think we have abandoned ritual - something to be found only in churches, synagogues and mosques. Not so. Secular life is suffused with many ritualistic activities, science communication among them. Science communication is continually affirming the value and importance of science. To see it only in terms of its manifest function in achieving certain outcomes is, as Merton reminds us, to miss something important.

5. THE PHILOSOPHY

What we miss, if we focus on communication as cause and effect, is that there are other ways of thinking about communication that lead to different explanations of what is happening, different expectations, and importantly different ways of approaching communication.

In this section we will discuss two alternatives, contrasting them both with the cause and effect view. First we will discuss the idea that communication is about following shared rules, but this turns out to be just a variant of the cause and effect view. Second we will discuss the idea that communication is formed through conversations, and this idea gives us a way through into more productive work.

5.1 Following the Rules

Take a simple and often used example: a game of chess. You can understand a chess game and take part in a game only if you know the rules. If you do not know the rules, you cannot make sense of what the players are doing, or why. Martian scientists might try to explain what is happening by suggesting that the chess pieces cause the players to act in certain ways, but this would not get them very far. The point to understand here is that there is, at first sight, something quite different about explaining something in terms of cause and effect and explaining something in terms of rules of conduct.

The idea that communication is concerned with following rules has been very popular in the last few decades. Several influential ways of thinking are based on this idea: transformational grammar, semiotics, and cultural studies among them. There are many popular accounts of this way of thinking (e.g., Campbell, 1982).

Explanations of communication based on this idea argue that all communication is based on codes - underlying rules and conventions that are used first to construct and then to interpret messages. A communicator encodes a message using the rules, and the receiver decodes the message using the same rules. Once we know what the rules are, so the explanation runs, we can construct effective communication.

This has become a popular way of describing communication because it resonates so well with contemporary information technologies and with research on genetic codes. Moreover, it seems so simple. Unfortunately, it is not simple. In order for the idea to work we need to also explain how rules are created, stored and changed. For information technology, this is fairly clear and straightforward; the rules are written for anyone to see. For genetics, it is more difficult and involves much speculative biology and philosophy; but treating genetics as a code does have some explanatory and predictive value.

But for human communication, such an idea is a logical absurdity. The rules (if they exist) are never visible and can only be inferred from observing people engaged in talking, reading, watching, singing, winking, waving and myriad other communicative activities. This leads us to a circular argument: these people are communicating in a certain way, therefore a rule to behave in that way must exist; we know the rule exists because people are behaving according to that rule. But even if this logical absurdity were unimportant, there is even more disturbing news for those who want to use such an idea in practice: it does not lead to any predictable outcomes. All that complicated explanation for no return, except perhaps the illusion that somehow one has understood how communication works.

The final weakness of the encoding/decoding idea is that it turns out to be just a variant of the cause-effect explanation. At the heart of any encoding or decoding - whether human, electronic or genetic - is a causal chain of events. What looks like a different type of explanation of communication turns out to be just a variant of the same thing. The only difference is that it emphasises an aspect of the process - encoding and decoding - rather than the overall causal relation between senders and receivers of a message.

5.2 Taking Part in a Conversation

Another, more promising approach, starts from a totally different idea, namely that all types of communication are variants of conversations between people. In this explanation there is no need to posit hidden rules or cause-effect relations in order to explain communication.

Communication is an overt phenomenon. It occurs, publicly or privately, between things: between people and people, people and deities, people and exhibits, people and books, people and pets, people and television, people and their garden sheds or jigsaw puzzles or fishing rods; in fact, it occurs between people and whatever has significance for them. It is a dynamic activity, which means that it changes and evolves as it unfolds. Because it occurs 'between', and is entirely dependent on the 'between' relationship, it cannot be reduced to its components. Thus there is no such thing as a message on its own, no such thing as a communicator on their own, no such thing as an audience on its own. We can only talk about what goes on between communicators and messages, audiences and messages. And because it is something that 'goes on' in a particular context over a particular period of time, we cannot talk about it as something static.

This idea is a profound break with most ways of thinking about communication. Scientists will be familiar with the central philosophical idea. In science, for example, the differences between biology and ecology as fields of study are profound. But the differences are not to do with subject matter. Both subjects study living things. The difference is in point of view and the units of observation and analysis. Moreover, and this is the important point, ecological factors, which are relational, dynamic, and changing, are not reducible to biological factors which are not. Similarly, in the conversational view of communication, we and others have argued that previous research in communication has failed because it has always tried to reduce the process to its component parts on the assumption that there is no loss of either data or explanatory power. But there is. Reducing communication to psychological or social factors misses the point of what is going on.

6. THE PRACTICE

What, then, should science communicators be doing?

Starting from the point of view that communication is dynamic, that it is an interactive, relational activity that goes on between people, then the only way to study communication is to observe a communicative activity as it is going on. Suppose one wanted to find out about the effectiveness of a science centre exhibition. The normal approach is to employ certain conventional methods, post-attendance surveys and focus groups being the most widely-used. However, the process of communication cannot be reduced to people's recollection of the experience, because by then, no activity is actually happening. The only useful way to investigate it is *in situ*, to look at the process as it happens, as it unfolds.

Thus direct observation is the most important research method to be used in communication research. If you want to know what is going on, you have to watch it actually going on. And if you want certain things to happen, you have to be there to see if these things are happening. Moreover, you can not only see what people are doing but talk to them about what they are doing as they are doing it.

Because communication between things is something that changes as it happens, according to immediate context; and because this type of observation is not neutral and objective in the traditional 'scientific' sense, but is participatory and involving; so a science communicator is in an ideal position to be not merely an observer and surveyor but an active participant and facilitator. But this entails two very important conditions.

The first condition is that the science communicator must have a good idea about what is to count as the success or failure of the science object. The second condition is that the science communicator must always remember that communication is not a matter of 'getting the message through' to passive learners. Participating in an observation study with an active audience must not be confused with telling people what to do and think. Watch what people are doing, see if they are doing it in the way you'd like them to, and if they are not, work out (from watching their behaviour) what you can do to make them more likely to do it.

Communication Research Institute of Australia, PO Box 398, Hawker, ACT 2614 Australia

7. REFERENCES

Blood, R.W. (1989) Public agendas and media agendas: Some news that may matter, *Media Information Australia*, 52, 7–15.

Campbell J. (1982). *Grammatical Man*. Harmondsworth: Penguin.

Klapper, J. T. (1960) *The Effects of Mass Communication*. Glencoe, Illinois: The Free Press.

McGuire, W. J. (1986). *The myth of massive media impact: savagings and salvagings, Public Communication and Behaviour, 1*. Orlando: Academic Press, 173-257.

McQuail, D. (1987). *Mass Communication Theory: An Introduction*, (Second Edition). Beverly Hills: Sage.

Merton, R. K. (1957). *Social Theory and Social Structure*. New York, The Free Press, 1949; enlarged editions, 1957, 1968.

Schudson, M. (1984). *Advertising: the Uneasy Persuasion*. New York: Basic Books.

L.J. RENNIE

7. COMMUNICATING SCIENCE THROUGH INTERACTIVE SCIENCE CENTRES: A RESEARCH PERSPECTIVE

It is 3.30 pm on Thursday as I pass through the entrance to *Scitech* Discovery Centre. "Hello", I say to the young person at the door, "It's the Education Committee meeting today!" She smiles and waves me on. The entrance to *Special Effects II*, the current major exhibition, looms large before me. Side-stepping the opening and weaving between the colourful scent exhibits of the *Perfumed Garden*, I head towards the office area at the back. The Centre is relatively quiet now, school groups have gone, a few adults wander about. One, thoughtfully and deliberately, touches each of several different materials set into a large toilet seat, apparently comparing the feeling of hotness or coldness resulting from different heat conductivities. Two small children press their faces to the glass wall of the chicken hatchery, and a woman watching them from a seat nearby loosens her shoes with a sigh. A blast of sound hits me: "Melanie! Melanie! Where's Daddy?" I realise I have walked between the *Whispering Dishes* and move aside. Tiny Melanie, held aloft by a Sciguide so that her ear reaches the focal point of one Whispering Dish, looks around to find her father waving from 40 metres away at the other. She looks surprised, confused, then begins to laugh.

Bright, cheerful, usually busy, mostly noisy, and nothing like an orderly science laboratory, yet this science centre claims to be a purveyor of science. Indeed, increasing the public's interest and participation in science is central to its stated mission. But is it likely to be successful? Almost since their inception, science centres have had their detractors who claim that science is of secondary consideration, sacrificed to the primary purpose of public amusement. Champagne (1975), Fara (1994), Parkyn (1993), Ravest (1993), Shortland (1987) and Wymer (1991) are some who have suggested that education loses out in the quest to entertain. Is it contradictory to have science and entertainment in the same place? Do science centres *really* communicate anything about science? Or do visitors just have fun and go away with their awareness of science unchanged?

Answers to questions like these must come from research; anecdotal evidence is not sufficient. The body of research in science centres and science museums is now considerable but, even though much has been learned, the answers are not as complete and clear-cut as we might wish. This chapter reports on this research with a particular focus on what it means for the communication of science through interactive science centres. It begins with an overview of the nature of research in places generically called museums, including science centres, zoos, art galleries, environmental centres and other similar institutions, followed by a synthesis of

S.M. Stocklmayer et al. (eds.), Science Communication in Theory and Practice, 107–121.
© 2001 *Kluwer Academic Publishers. Printed in the Netherlands.*

findings for interactive science centres. A discussion of these findings leads to a reappraisal of the research questions we need to ask in order to generate the kinds of information science centres can employ to enhance the communication of science to their visitors.

1. RESEARCH IN MUSEUMS: AN OVERVIEW

Traditionally, research in museums has focused on answering questions about either the exhibits or the visitors. Questions about exhibits are usually evaluative, aimed at exhibit appraisal. Questions about visitors have several foci, including why visitors come to museums, what they do and what they learn. In both kinds of research, the data are collected from the visitors, because the effectiveness of an exhibit (and indeed the whole museum) must be measured in terms of how visitors react to it. The broad term 'visitor studies' might be used to describe research in which these data are collected and, as shown diagrammatically in Figure 1, there are three main purposes for this kind of research. Visitor studies may be designed to find out about the visitors themselves, including demographic and some times psychological characteristics, to prepare what Pearce and Moscardo (1985) call visitor profiles. Another focus of research is the investigation of visitor behaviour, which Pearce and Moscardo call visitor reaction. These authors also mention visitor impact, which relates to monitoring the physical effects of wear and tear that visitors have on the exhibit itself and its location. Museums and science centres usually monitor the demographic profiles of their own visitors and visitor impact is attested to by maintenance staff but neither profiles nor impact are the concern of this chapter. Here the focus is on visitor behaviour - how visitors use the museum environment, their interaction with exhibits and what the science-related outcomes of their visit might be.

Figure 1 shows that research into visitor behaviour has two main purposes. One kind of research attempts to learn about what visitors actually do during their visit to the museum, including their responses to the physical design and features of the museum environment and their participation in social interactions. Researchers have mapped the paths visitors follow through the museum, and recorded how much time they spend looking at exhibits and doing other things, such as visiting the cafeteria. Hein (1998) provides examples of a number of tracking methods and studies. Social factors, such as the size, composition and characteristics of visiting groups and their interactions have also been well studied. One particular line of research into social factors concerns family behaviour during museum visits (see, for example, Diamond, 1986; Dierking & Falk, 1994; Hilke & Balling, 1985; Kropf, 1989; McManus, 1994).

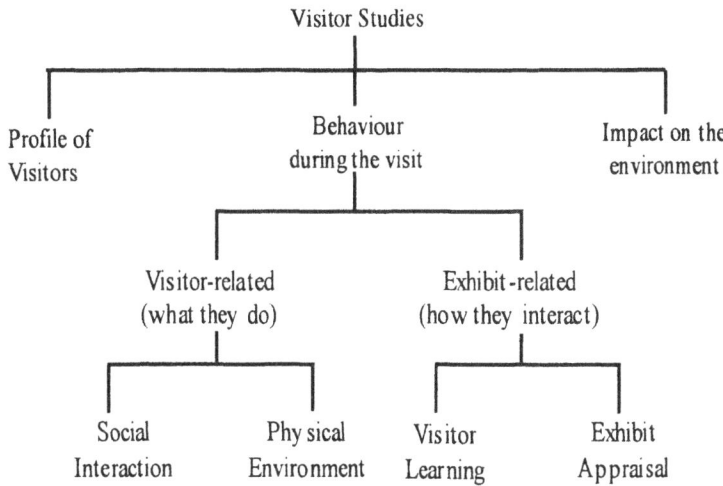

Figure 1. Some foci of research in museums and similar institutions

The purpose of the second kind of research into visitor behaviour concerns their reaction to, or interaction with, particular exhibits. Much of the early work in this area related to exhibit appraisal, and terms such as the attracting and holding power of exhibits were introduced to measure how successful they were at engaging visitors' attention. Examples include research evaluating exhibit design and the features associated with its effectiveness, such as the format and positioning of labelling. Screven (1984) provides a bibliography of evaluation in museums. Other research about visitors' reactions to exhibits has examined the effect on the visitor and the learning process. Besides documenting visitors' interactions with museum exhibits, this research has attempted to relate cognitive and affective learning outcomes to the nature and design of the exhibits and to the effects of other variables, such as interaction between visitors and explainers, and pre- and post-visit instruction. Some studies have focused on how people actually learn science through interacting with exhibits. For example, Feher and her colleagues (Feher, 1990; Feher & Diamond, 1990; Feher & Meyer, 1992, Feher & Rice, 1985) have studied the nature of the learning process and Borun (Borun, 1989; Borun, Massey & Lutter, 1993) has investigated the naive conceptions visitors bring to the exhibit.

2. RESEARCH INTO VISITOR LEARNING

A feature which dominates all research into the activities of visitors to museums, including science centres, is that these venues are free-choice settings, in other words, they are places where visitors can please themselves about what they do. They have opportunities for social and cooperative learning and are able to investigate things in a relatively unhindered way. They can determine their own levels of participation, and their own ways of interacting. The learning situation is quite different from that in formal places of learning, such as schools, because museum 'tasks' are non-evaluative, non-competitive and people can choose whether or not to engage in them (Falk, Koran, & Dierking, 1986). Further, visits are short, perhaps two hours in a year, and visits to particular exhibits are fleeting. Can we expect learning to occur in such a short time? And yet, the exhibits are designed to offer visitors the opportunity to learn from their experiences. In their paper on a technology for effective science exhibits, originally published 20 years ago, Miles and Tout (1994) describe five educational aspects of exhibits, all designed to promote visitor learning. The potential for visitors taking away something worthwhile from their visit is high but capturing that elusive 'something' through research is not easy. There are several reasons for this.

First, it is clear that each visitor has a unique experience and as a consequence, the outcomes are likely to be unique. Not only do they choose which exhibits to visit, how to interact with them, and for how long, visitors bring with them their own unique blend of knowledge, experience and expectations. Falk and Dierking (1992) refer to this as the personal context of the visit experience and it plays a major role in determining the nature and outcomes of the visit. For example, Tulley and Lucas (1991) report research on a *Lock and Key* exhibit which challenged the visitor to assemble it. They found that males and females had different approaches to the assembly task and those visitors with experience with locks took much less time. Thus, the cognitive complexity of the interaction with an exhibit will depend on the match between the visitor's background knowledge and the cognitive level of the exhibit. Another factor that contributes to the uniqueness of the visit experience is the social context of the visit, concerned with who accompanies visitors, the people they meet and how they interact socially. Visits with family and friends provide opportunities for discussion, joint experimentation, and often visitors 'tutor' one another (Diamond, 1986; Tuckey, 1992). In addition to the personal and social contexts of the visit, there is the physical context comprising the exhibits themselves and the surrounding museum environment. Falk and Dierking (1992) point out that the nature of the visit experience is determined by the interaction between these three contexts. A comprehensive research approach must take account of this interaction in order to capture the unique visit experience.

A second complicating factor in visitor research is the difficulty of collecting data. Lucas and his colleagues (Lucas, 1983; Lucas, McManus & Thomas, 1986) have documented a number of these problems. One problem relates to collecting data in ways that do not affect how the visitor behaves. Unobtrusive observation, or recording with video or audiotapes are productive methods of data collection but there are ethical problems if visitors are unaware of this surveillance, and problems of their reactivity if they are aware. Intrusive methods of data collection, such as interviewing visitors about an exhibit as they stand before it, or using photographs or videotapes to help them recall their interaction with it, are good ways of finding out what visitors understand about an exhibit. However, the researcher must be aware that the process of the data collection is likely to change the nature of the experience. The results will describe what the visitor knows after the reflective experience of the interview but this may not be the same as the outcome produced by an uninterrupted visit. For example, Allen (1997) used a number of thought provoking activities in combination with structured interviews to examine visitors' understanding of a 'coloured shadows' exhibit. Her purpose was not to investigate the outcome from the visit experience but to find out which activities facilitated visitors' understanding so that the exhibit might be improved. Appropriately, Allen noted that her research design was effective in promoting understanding of visitors' reasoning patterns but that it changed the nature of the visit experience. To measure the outcomes of an uninterrupted visit requires a less intrusive approach.

Yet another factor adds complexity to the research task. Because learning is a cumulative process, the visit experience cannot be regarded as an isolated event. Learning involves the linking of new information into a person's existing mental structures in ways that enables it to be used at other times. Thus learning might 'happen' during the visit, when ideas and information suddenly fall into place with dramatic clarity, and the visitor experiences what Friedman (1995) calls a "tiny epiphany'. But sometimes that 'Ah ha!' experience might happen later, when remembered aspects of the visit experience unexpectedly come together as a result of a separate incident. Clearly, follow up studies would be necessary for the researcher to take account of this longer term effect of the visit.

Capturing the elusive learning experience, describing its nature and pinning it down to a particular sequence of events is not accomplished easily using traditional research approaches. Pre-test/post-test, control group studies have limited success when the subjects' experiences are all unique and confounding variables cannot be controlled. Thus the advantage gained by being able to compare before and after measures to assess the impact of a visit has to be balanced against the decontextualising of the visit experience in order to obtain a controlled research environment. Gradually, researchers have moved away from these traditional designs and many researchers would argue that firmer findings have resulted from small-sample, interpretative or naturalistic approaches. Hein (1998) provides an historical overview of the research approaches used in visitor studies including a detailed discussion and comparison of the different research methods. Falk (2001) explores some of the current knowledge about, and attempts to establish a research agenda, for free-choice science education, including in the museum context.

It is important to note that learning, as discussed in the preceding paragraphs, is not restricted to cognition. Friedman's (1995) epiphanies (a term also used by Hein, 1998) are much more than cognitive events, as all who have experienced them can testify. Learning in the context of museums is considered to be a much broader term. I have found this definition useful:

> learning is an internal change in a person - the formation of new mental associations or the potential for new responses - that comes about as a result of experience. (after Woolfolk, 1987)

Thus learning might be cognitive, or affective, or psychomotor, or social, or any combination of these, but for any learning to occur, some mental or physical activity is required. Learning is also personal and internal, but it may be observed through the learner's changed behaviour. In the museum context, mental or physical activity relating to the exhibits is a learning experience if it results in the formation of new mental associations or the potential for new responses. If the exhibit is science-related, then that learning may be associated with the communication of science.

3. VISITOR RESEARCH IN INTERACTIVE SCIENCE CENTRES

The term 'science centre' is used here rather loosely, to refer to collections of exhibits that are designed to represent particular science concepts. This is in contrast to the exhibits in traditional museums, whose purpose is linked to the intrinsic nature or value of the object, and which portray features which are properties of the objects themselves. Of course, the visitor plays a part in making meaning from the object (Gurian, 1999). For example, a dinner plate may appear to be quite ordinary, but if it were salvaged from the *Titanic*, it assumes a significance depending on the visitor's knowledge of the *Titanic*. In contrast, exhibits built to present concepts or ideas may be quite abstract and the ideas illustrated often depend on interaction, or even interference, from the visitor to make their point. Collections of this kind of exhibit might be in purpose-built interactive science centres, like *Scitech*, the subject of the opening paragraph, or be a small part in a larger museum, such as the well known *Launch Pad* in London's Science Museum. Science exhibits described as interactive are those that are meant to provoke action from the visitors, provide responsive feedback and thus provoke further interaction. They may be 'hands-on' because they rely on the physical involvement of the visitor or they may require other senses to perceive their message. In essence, they are dynamic, rather than static, requiring active participation by the visitor. Many exhibits represent concepts relating to technology rather than science (such as an exhibit demonstrating conservation of energy and cost savings by using fluorescent rather than incandescent lighting), but for the purpose of this review, no distinction is made.

3.1. What Are the Outcomes from Science Centre Visits?

More than two decades ago, in a discussion of the learning potential of science centres, Kimche (1978) recognised the need for visitor research to go beyond demographics and the popularity of exhibits and to find new ways of measuring the educational outcomes for visitors to science centres. Progress has been slow, as several reviews reveal (Bitgood, Serrell & Thomson, 1994; Ramey-Gassert, 1997; Rennie & McClafferty, 1996). This is partly because of the reluctance of researchers to look at a range of outcomes besides cognitive learning, but also, as noted earlier, because of the difficulties in designing research to identify and measure these outcomes. Lucas and his colleagues (Lucas, 1983; Lucas et al., 1986) describe and discuss a number of approaches in science museums, and Hein's (1998) summary of learning in museums gives illustrated examples of a number of techniques for collecting data. In their review of science learning in science centres, Rennie and McClafferty (1996) summarise the findings of a range of approaches to measuring learning, including observation, written measures, open-ended questions and interviews, stimulated recall, and the more traditional comparison studies. But the variety of approaches to research, and the variety of questions to which that research sought answers, have resulted in a volume of findings which, although extensive, are piecemeal and difficult to synthesise. Rennie and McClafferty concluded their review by stating

> There emerged a picture consistent at only the most general level, which suggests that some cognitive, affective and psychomotor learning occurs most of the time, but there is considerable variation across science centres and also across exhibits within centres. (p. 86)

Rennie and McClafferty's conclusion of consistency at only a general level is not surprising, given the variations in the personal and social contexts of each visitor's experience, as was discussed earlier, as well as variation in the physical contexts of different science centres and exhibits.

3.2. What Kinds of Exhibits Are Successful?

Let us put aside the personal and social contexts for the moment and focus on the physical context, because this is the one over which science centres have most control. What does research tell us about the kinds of exhibits that are most likely to promote visitor engagement and learning? Perhaps the most comprehensive work on successful interactive science exhibits and visitor learning has been done by Perry (1989). She considered the success of an exhibit in terms of two conditions: visitors had to enjoy themselves and they had to learn something. Perry reviewed an extensive literature in psychology, motivational theory and instructional design, to develop, then test and refine a model to describe an intrinsically motivating exhibit. The model was later retested and validated by Peiffer (1995) in another science centre. Perry's (1989) model had six summary criteria:

Curiosity: the exhibit should surprise and intrigue the visitor;
Confidence: the exhibit should promote perceptions of personal competence;
Challenge: the exhibit should challenge the visitor, by maintaining a balance between success and effort;
Control: the exhibit should promote feelings of self-determination and control;
Play: the exhibit should promote feelings of enjoyment and playfulness; and
Communication: the exhibit should stimulate meaningful interaction.

Perry (1992) used her model to demonstrate how motivating exhibits could be designed, and extended it to describe the successful museum visit (Perry, 1993). Using her six criteria, Perry argued that visitors' needs are met when they participate in intellectual, social and physical interactions, when they learn something new, or become more aware of, or interested in, something. If science centres are to be successful in communicating science, that something must be related to science.

4. WHAT IS MEANT BY THE COMMUNICATION OF SCIENCE?

Nine years ago, McManus (1992) closed a review of education in science museums with a call for more work on "the assessment and description of the nature of their communication activities" (p.180). An extensive review of the impact on learning of visitors to museums by Bitgood, Serrell and Thompson (1994) cited over 200 references, but found few attempts to measure the nature of learning, and none on the communication of science. In 1996, Rennie and McClafferty concluded that McManus' plea was still current. In most of the studies they reviewed researchers aimed to measure increases in science knowledge and sometimes attitudes. A number looked at memories (Stevenson, 1991, is a good example), revealing a wide range of visitor responses in learning science, but there was little attention paid to the process of communication. A notable exception is a paper by Lucas (1983), who extended the discussion beyond science knowledge by examining the relationship between scientific literacy and informal learning. As a basis for his analysis, Lucas used Shen's (1975) tripartite perspective of practical, civic and cultural scientific literacy. Science centres and museums were just one part of Lucas' review, and it is surprising that no researchers seem to have picked up these ideas and tested them more fully by research. Lucas' (1983) paper remains one of the more comprehensive analyses of the perspectives about science, rather than just the knowledge of science, which may be communicated by science centres.

Despite the halting progress, we can say that we have accumulated through research some useful knowledge about visitors' learning and the kinds of visit experiences that are likely to result in learning about science. First, people do learn in interactive science centres and quite dramatic learning events can occur, as research work, particularly into memories, has revealed. Second, learning is difficult but not impossible to measure. Third, the outcomes for the visitor may include cognition, affect, social or psychomotor skills. Fourth, we have identified the kinds of exhibits which are likely to promote interaction and therefore learning.

These findings are not unique to science centres, similar summaries have been provided for museums in general by Falk, Dierking and Holland (1995) and Hein (1998). How can these findings illuminate the purpose of this chapter, which is concerned with the communication of science through interaction with exhibits in science centres? The answer depends on what we understand by 'the communication of science'. If we hope that a deep cognitive understanding of science concepts results from every science centre visit, we are likely to be disappointed. If we hope that people have a science-related encounter which enables them to make more sense, in a scientific way, of their experiences, the research suggests that our hopes are likely to be fulfilled. But what is meant by 'making sense in a scientific way'? Here are two perspectives, both from a book nicely titled *Sharing Science* (Quin, 1989). One is by Richard Gregory, a neuropsychologist, who at the time was Science Director of The Exploratory. The other is by Jerry Wellington, a leading science educator.

In his keynote essay, Gregory (1989) wrote about hands-on, hand-waving, and handle-turning to represent different levels of understanding in science. 'Hands-on' has its usual meaning of perceptual exploration, and Gregory argues that

> although hands-on experience is effective, indeed essential, for learning to see objects - hands-on experience can hardly be adequate for arriving at scientific understanding. (p. 3)

It is the interpretation of the perceptual exploration which creates meaning. Gregory describes as 'hand-waving explanations', the intuitive or commonsense meaning-making which allows 'phenomena and principles to be richly understood' (p. 3). Although these explanations are not logically structured or quantified, Gregory believes they may be necessary for the more formal logical or mathematical interpretations of phenomena to have meaning. These are the 'handle-turning"' or computational accounts of science embodied in equations such as $E=mc^2$, which has little intuitive meaning on its own. It is the hand-waving explanations that give the useful intuitive accounts that Gregory argues should be the major aim of interactive science centres, for two reasons. First, these give meaning to the perceptual, hands-on experience, and second, they underlie the rigorous handle-turning scientific methods and mathematics which are the important bases for science and technology.

The second perspective comes from Wellington (1989), who argues that science centres contribute to science education through each of the cognitive, affective and psychomotor domains familiar to educators. A fundamental premise is that these domains are not separate but inter-related. Experiences in the affective domain – interest, motivation, eagerness to learn–and those in the psychomotor domain – manipulative skills, hand-eye coordination–contribute to knowledge and learning in the cognitive domain. Wellington describes knowledge in terms of knowledge *that*, knowledge *how*, and knowledge *why*, suggesting that in science centres knowledge *that* predominates almost exclusively over knowledge of *how* and *why* particular phenomena occur. But, continues Wellington,

> while hands-on science centres may not contribute immediately and directly to deep understanding, their *indirect* effect must not be forgotten (p 31, original emphasis).

The argument here is that the science centre experience can provide the ground work for later and deeper understanding.

Perhaps an example will clarify these perspectives. In my role as researcher during one science centre visit, I watched a woman gazing thoughtfully at the *Pulleys and Weights* exhibit. Tentatively at first, then more purposefully and with a slight grimace, she pulled the rope equipped with a single pulley to lift a 15kg weight. Carefully, she let down the weight and then, after a pause, pulled the next rope to use two pulleys to lift a second 15kg weight. Smiling slightly, she moved quickly to the third rope and easily lifted its 15kg weight using three pulleys. She stood back smiling. I stepped forward, inquiring whether I could ask her some questions about the exhibit. 'Pulleys!' she said, not waiting for my questions. 'I was listening to a man in a garage a couple of weeks ago, and he said it was easier to lift things when you used pulleys, and the more pulleys you had, the easier it was to lift!' She beamed at me, clearly delighted with her experience.

From Gregory's (1989) perspective, this woman's perceptual exploration of the exhibit was a hands-on experience that led to her 'tiny epiphany' about pulleys and weights. She recalled a past experience, which had no meaning to her at the time, and developed a hand-waving explanation of the experience. The relationship between effort and the number of pulleys was now obvious to her in an intuitive and sensible way. It is doubtful that she could express this relationship in a mathematical statement, or that she could explain why the relationship was inverse, so it is unlikely she had a handle-turning level of understanding. But clearly, the woman had enjoyed a learning experience, and she made sense from that experience in a scientific way. From Wellington's (1989) perspective, this woman's interaction with the exhibit was a 'knowing that' experience, but one that involved all three educational domains, cognitive, affective, and psychomotor. Actually pulling the rope was a psychomotor learning experience, comparing the feel of the same weights with different numbers of pulleys involved cognition. Her smiles and her eagerness to tell me, a stranger, what she had learned typify the affective, motivating reaction through which Wellington believes science centres can make a significant contribution. I do not know whether, later on, the woman had an opportunity to find out the knowledge why, or how, that accompany deep understanding, but it seemed to me that such an opportunity would be seized with enthusiasm. Using both Gregory and Wellington's perspectives, the interaction would be interpreted as demonstrating that learning has occurred, and scientific meaning was communicated.

5. SCIENCE CENTRES AND THE COMMUNICATION OF SCIENCE

The potential role of science centres in the communication of science is increasingly important for at least three reasons. First, and simply, science centres are proliferating – as there are more of them, they offer science to a larger potential audience. Second, they provide opportunities to engage with science to the majority of the adult population who are no longer part of a formal educational system which offers science as part of its curriculum. Third, perhaps tragically, and despite our best intentions, schooling usually has not been successful in creating scientifically literate school leavers. Shamos (1995) devotes his book, *The Myth of Scientific Literacy*, to clarifying the purpose of science education, the meaning of scientific literacy, and the reasons for past failure to achieve it. Significantly he points to science museums as "probably the most logical sites of adult literacy or awareness"

(p. 187) in science. From the perspective of a science educator, Fensham (1997) examines the failure of schools to achieve scientific literacy in all their students by analysing the changes in science curricula, and the reasons for their apparent lack of success. Fensham concludes that the major problem is pressure from the power players involved, including academic scientists, for a continued focus on science education as preparation for further studies in science, in other words, a focus on the content knowledge of science. Shamos (1995), too, argues for a science education which moves, among other things, away from learning science content and towards teaching science in order to develop appreciation and awareness of the scientific enterprise. Instead of knowing enough science to make independent decisions, as popular definitions of civic scientific literacy would imply, Shamos argues that it is more important to 'emphasise the proper use of scientific experts' (p. 217). So, in Shamos' terms, 'what we seek is a society that (a) is aware of how and why the scientific enterprise works and of its own role in that activity, and (b) feels more comfortable than it presently does with science and technology'(p. 219).

If we are to accept these points of view, even partially, then we must acknowledge that whatever science content knowledge is communicated, it can be only a tiny part of the communication of science by science centres. A much larger part of what is communicated, both implicitly and explicitly, is about the nature of the scientific enterprise and the role society plays in it. We have seen that much of the research has tried to measure what knowledge people take away from a visit, yet that is hardly likely to be the important question. Knowledge of facts slips the memory quickly as the advancing years remind us, sometimes painfully. Further, considerable and very thorough research (Layton, 1991; Jenkins, 1994; Wynne, 1991) has established clearly that people do not absorb scientific knowledge, unchanged, from any source. Rather they make meaning from it in their own context, translating and reworking it into something they can use in their own specific circumstances. Gregory's (1989) term, hand-waving explanations, is a good description of what people understand from the knowledge they have obtained. Layton, Jenkins, Macgill, and Davey (1993) give examples of people's explanations, some of which are quite inaccurate, yet they use them in their daily decision-making.

Such evidence suggests, yet again, that asking whether people learn science from a visit to a science centre is the wrong question. Communicating science isn't just about learning science content, it is about developing "long-term relationships with the content, phenomena and issues of science" (St John & Perry, 1993, p. 59). In a report prepared for ASTC, St John and Perry (1993) urge

> science museums to see themselves as institutional resources forming part of the nation's
> education infrastructure. (p. 59)

In this context, questioning whether science centres contribute to long-term relationships with science is more appropriate if our research questions are to ask about how and whether science centres communicate science.

A final *caveat*: we must not forget the criticisms levelled at science centres, because they are not routinely addressed. Champagne's (1975) criticisms of the science centre he visited with his family are especially pertinent, not only because they have not been refuted fully, but because if science centres are to communicate an awareness of science, the nature of that science is critical. Champagne accused the science centre of presenting poor science on a number of counts. First, the focus on exciting demonstrations obscures what science is really about: the asking and answering of questions about how the world works. Second, 'sloppy' science was presented by careless definitions and explanations, and by exhibits which combined several variables so that there was no way to sort out their independent effects and answer clearly questions about how they work. Third, no attention was paid to the ethical dimensions of the scientific and technological decisions made by humans. Fourth, science was portrayed as easy and unproblematic when instead "its human fallibility as well as its attempts toward integrity to achieve its results" (p. 39) should be portrayed. Champagne has not been alone in making these kinds of criticism, Shortland (1987) and, more recently, Parkyn (1993) have conveyed similar concerns.

How are science centres to portray more realistically the process of science? Products (especially technological products) are easy to present as exhibits (many science centres have a cut away engine with a handle to turn the pistons), and concepts, too, are relatively easy (the ubiquitous Bernoulli ball). But too often these are displayed without human contexts. Lucas (1983) concluded that most learning takes place when the material is placed in context. How can the context of the exhibits be shown more effectively? How are the social aspects of science to be portrayed? Arnold (1996) describes some of the exhibition techniques being used by museums to sit science in a social context, ranging from actors dramatising the work of scientists, to computer games, to direct material evidence of controversy. Strategies such as these are yet to be evaluated, especially in terms of the image of science and technology portrayed. As Schauble and Bartlett (1997) point out, change and redesign are more successful in a research-museum partnership. Thus, however science centres contrive to enhance their ability to communicate science, the questions about their effectiveness will need to be answered by research.

What is the match between science and the cheerful, noisy, active science centre environments, with floor staff (explainers) in their colourful jackets - hardly the image we have of scientists working in science laboratories. How can such places tell us anything about real science and the scientific enterprise? If we limit our view of the science and scientific enterprise to stereotyped views of the idealised scientist in his (rarely her) laboratory, then probably not much. But real science isn't like that, although Petkova and Boyadjieva (1994) have argued how the idealised image functions to preserve the separation of science and scientists from other social communities. Further, science centres, realistically, cannot aim to turn all of their visitors into scientists, only people who might know a bit more about science and be more aware of it in their daily lives. And this aim is one which can be achieved by good science centre experiences.

Curtin University of Technology, Perth, Western Australia

6. REFERENCES

Allen, S. (1997).Using scientific inquiry activities in exhibit explanations. *Science Education, 81,* 715-734.

Arnold, K. (1996). Presenting science as product or as process: Museums and the making of science. In S. M. Pearce (Ed.), *Exploring science in museums.* London: the Althone Press, 53-78.

Bitgood, S., Serrell, B., & Thompson, D. (1994). The impact of informal education on visitors to museums. In V. Crane, H. Nicholson, M. Chen, & S. Bitgood (Eds.), *Informal science learning: What research says about television, science museums, and community-based projects.* Dedham, MA: Research Communication Ltd, 61-106.

Borun, M. (1989). Naive notions and the design of science museum exhibits. *Journal of Museum Education, 14* (2), 16-17.

Borun, M. Massey, C., & Lutter, T. (1993). Naive knowledge and the design of science centre museums exhibits. *Curator, 36,* 201-219.

Champagne, D.W. (1975). The Ontario Science Center in Toronto: Some impressions and some questions. *Educational Technology, 15*(8), 36-39.

Diamond, J. (1986). The behaviour of family groups in science museums. *Curator, 29,* 139-154.

Dierking, L.D., & Falk, J.H. (1994). Family behaviour and learning in informal settings: A review of the research. *Science Education, 78,* 57-72.

Falk, J.H. (Ed.) (2001). *Free-choice science education: How we learn science outside of school.* New York: Teachers College Press.

Falk, J.H. & Dierking, L.D. (1992). *The museum experience.* Washington: Whalesback Books.

Falk, J.H., Dierking, L.D., & Holland, D.G. (1995). What do we think people learn in museums? In J.H. Falk & L.D. Dierking (Eds.), *Public institutions for personal learning: Establishing a research agenda.* Washington, DC: American Association of Museums, Technical Information Service, 17-22.

Falk, J.H., Koran, J.J. Jr., & Dierking, L.D. (1986). The things of science: Assessing the learning potential of science museums. *Science Education, 70,* 503-508.

Fara, P. (1994). Understanding science museums. *Museums Journal, 94*(12), 25.

Feher, E. & Meyer, K. R. (1992). Children's concept of colour. *Journal of Research in Science Teaching, 29,* 505-520.

Feher, E. & Rice, K. (1985). Development of scientific concepts through the use of interactive exhibits in a museum. *Curator, 28*(1), 35-46.

Feher, E. (1990). Interactive museum exhibits as tools for learning: Exploration with light. *International Journal of Science Education, 12,* 35-39.

Fensham, P.J. (1997). School science and its problems with scientific literacy. In R. Levinson & J. Thomas (Eds.), *Science today: Problem or crisis?* London: Routledge. 119-236.

Feher, E., & Diamond, J. (1990). Science centres as research laboratories. In B. Serrell (Ed.)., *What research says about learning in science museums.* Washington: Association of Science-Technology Centres, 26-28.

Friedman, A.J. (1995). Learning moments. Association of Science-Technology Centres Newsletter, May/June, 7,9.

Gregory, R. (1989). Turning minds on to science by hands-on exploration: the nature and potential of the hands-on museum. In M. Quin (Ed.), *Sharing science: Issues in the development of interactive science centres.* London, Nuffield Foundation on behalf of the Committee on the Public Understanding of Science (COPUS), 1-9.

Gurian, E.H. (1999). What is the meaning of this object? *Daedalus, 128* (3), 163-183.

Hein, G.E. (1998). *Learning in the museum.* London: Routledge.

Hilke, D.D., & Balling, J.D. (1985). The family as a learning system: An observational study of family behaviour in an information rich setting. In J. D. Balling, D.D. Hilke, J.D. Liversidge, E.A. Cornell & N.S. Perry (Eds.), *Role of the family in the promotion of science literacy.* (Final Report for National Science Foundation Grant No. SED-811-2927,. 60-104).

Jenkins, E. (1994) Public understanding of science and science education for action. *Journal of Curriculum Studies, 26,* 601-611.

Kimche, L. (1978). Science centres: A potential for learning. *Science, 199*, 270-273.

Kropf, M.B. (1989). The family museum experience: A review of the literature. *Journal of Museum Education, 14* (2), 5-8.

Layton, D. (1991) Science education and praxis: the relationship of school science to practical action. *Studies in Science Education, 19*, 43-79.

Layton, D., Jenkins, E., Macgill, S., & Davey, A. (1993). *Inarticulate science?* Nafferton, East Yorkshire: Studies in Education, Ltd.

Lucas, A. M. (1983). Scientific literacy and informal learning. *Studies in Science Education, 10*, 1-36.

Lucas, A.M., McManus, P., & Thomas, G. (1986). Investigating learning from informal sources: Listening to conversations and observing play in science museums. *European Journal of Science Education, 8*, 341-352.

McManus, P.M. (1992). Topics in museums and science education. *Studies in Science Education, 20*, 157-182.

McManus, P.M. (1994). Families in museums. In R. Miles & L. Zavala (Eds.), *Towards the museum of the future: New European perspectives.* London: Routledge.

Miles, R.S., & Tout, A.F. (1994). Outline of a technology for effective science exhibits. In E. Hooper-Greenhill (Ed.), *The educational role of the museum.* London: Routledge, 87-100.

Parkyn, M. (1993). Scientific imaging. *Museums Journal, 93*(10), 29-34.

Pearce, P.L. & Moscardo, G. (1985). Visitor evaluation: An appraisal of goals and techniques. *Evaluation Review, 9*, 281-306.

Peiffer, B. (1995). *Interactive science exhibits on colour concepts. Testing an educational model.* Paper presented at the annual conference of the National Association of Research in Science Teaching, San Francisco.

Perry, D.L. (1989). The creation and verification of a development model for the design of a museum exhibit. (Doctoral dissertation, Indiana University, 1989). *Dissertation Abstracts International, 50* (12), 3926. (University Microfilms Inc. No, 9012186)

Perry, D.L. (1992). Designing exhibits that motivate. *ASTC Newsletter,* March/April, 9-12.

Perry, D.L. (1993). Measuring learning with the knowledge hierarchy. *Visitor Studies: Theory, Research and Practice, 6*, 73-77.

Petkova, K., & Boyadjieva, P. (1994). The image of scientist and its functions. *Public Understanding of Science, 3*, 215-224.

Quin, M. (Ed.) (1989). *Sharing science: Issues in the development of interactive science centres.* London, Nuffield Foundation on behalf of the Committee on the Public Understanding of Science (COPUS).

Ramey-Gassert, L. (1997). Learning science beyond the classroom. *The Elementary School Journal, 97*, 433-450.

Ravest, J. (1993). Where is the science in science centres? *ECSITE Newsletter,* (Summer), 10-11.

Rennie, L.J., & McClafferty, T.P. (1996). Science centres and science learning. *Studies in Science Education, 27*, 53-98.

Schauble, L., & Bartlett, K. (1997). Constructing a science gallery for children and families: The role of research in an innovative design process. *Science Education, 81*, 781-793.

Screven, C.G. (1984). Educational evaluation and research in museums and public exhibits: A bibliography. *Curator, 27*, 147-165.

Shamos. B.M.H. (1995). *The myth of scientific literacy.* New Brunswick, NJ: Rutgers University Press.

Shen, B.S.P. (1975). Science literacy and the public understanding of science. In S.B. Day (Ed.), *Communication of scientific information.* Basel: Karger, 44-52.

Shortland, M. (1987). No business like show business. *Nature, 328*, 213-214.

St. John, M., & Perry, D. (1993). A framework for evaluation and research: Science, infrastructure and relationships. In S. Bicknell & G. Farmelo (Eds.), *Museums visitor studies in the 90s* London: Science Museum, 59-66.

Stevenson, J. (1991). The long-term impact of interactive exhibits. *International Journal of Science Education, 13*, 521-531.

Tuckey, C.J. (1992). Schoolchildren's reactions to an interactive science centre. *Curator, 35*, 28-38.

Tulley, A., & Lucas, A.M. (1991). Interacting with a science museum exhibit: Vicarious and direct experience and subsequent understanding. *International Journal of Science Education, 13*, 533-542.

Wellington, J. (1989). Attitudes before understanding: The contribution of interactive centres to science education. In M. Quin (Ed.), *Sharing science: Issues in the development of interactive science centres*

London, Nuffield Foundation on behalf of the Committee on the Public Understanding of Science (COPUS), 30-32.

Woolfolk, A. (1987). *Educational psychology* (3rd ed.). New York: Prentice Hall.

Wymer, P. (1991). Never mind the science, feel the experience. *New Scientist*, 5 October, 49.

Wynne, B. (1991) Knowledges in context, *Science, Technology, & Human Values,16*, 111-121.

J.K. GILBERT

8. TOWARDS A UNIFIED MODEL OF EDUCATION AND ENTERTAINMENT IN SCIENCE AND TECHNOLOGY CENTRES

1. WHAT ARE SCIENCE AND TECHNOLOGY CENTRES FOR?

1.1. Competing or Complementary Aims?

McManus (1992) described the evolution of museums concerned with science and technology as having passed through three 'generations' over the last several centuries. The most recent of these has focused on the provision of experiences and the representation of ideas, rather than the presentation of objects, through the medium of 'interactive exhibits'. These involve a visitor, either singly or with others, taking some action on the interactive exhibit in response to simple instructions that are usually presented in an adjacent text panel. The response produced often leads to a train of additional actions and responses. The phrase 'Science and Technology Centre' has come to mean a collection of interactive exhibits. The great success of such centres with the public (Thomas, 1994) is leading to a breakdown of distinctions between them and the notion of 'museum: object-based displays are being mixed with interactive exhibits which are relevant to their theme.

The rapid proliferation of interactive exhibits over the last decade has inevitably led to a questioning of their purpose. Many see them primarily as a source of entertainment - for example, Shortland (1987). The argument goes that the general public is uninterested in science and technology as not being 'fun'. Thus, if interactive exhibits provide 'fun', they will lead the public to a more general willingness to engage with science and technology. Others, such as Stevenson (1991) see them as a source of education. The argument here is that active engagement with interactive exhibits will lead both to a greater understanding of the ideas on which they are based and to an improved disposition towards science and technology.

These two views are often seen as being in direct competition with each other, perhaps because they have different protagonists within the staffs of science and technology centres. Those with an educational brief will perhaps place more emphasis on 'learning' whilst those with a managerial and financial brief will pay more attention to entertainment. A brief examination of the meaning of the two words in the Oxford Concise Dictionary (1959) shows this to be uncalled for. 'Entertain' is defined as

S.M. Stocklmayer et al. (eds.), Science Communication in Theory and Practice, 123–142.
© 2001 Kluwer Academic Publishers. Printed in the Netherlands.

> maintain (correspondence, discourse); arouse, occupy agreeably (person); receive
> hospitality; harbour, cherish, welcome, consider (idea, feeling, proposal)

The entertainment function of a science and technology centre is usually seen to involve actively welcoming visitors and in providing them with agreeable activities and valuing their responses. 'Educate' is defined as

> bring up (young person); give intellectual and moral training to; provide schooling in;
> train (person, oneself, a faculty) to do

The educational function of a science and technology centre might thus be seen to provide visitors with the opportunity to do some intellectually worthwhile activity, leading to an understanding of, or a 'schooling in', the underlying ideas. These two interpretations can be brought together, so that a centre could be, even should be, a place which actively welcomes visitors, providing them with activities which they enjoy and which are can lead an understanding of scientific and technological ideas. Such a perspective is consistent with the constructivist view of education, where learning is the result of an active engagement in the light of previous experience and knowledge (Kelly, 1963; Lave 1991).

One potential obstacle to an address to this combined function is the possibility of tension between the 'science' focus and the 'technology' focus of science and technology centres.

1.2. Science, Technology, or Science/Technology?

The issue here is about what gets represented in science and technology centres: is it science, or technology, and in what proportion? These questions can only be addressed with the use of a common method of analysis. One developed by Pacey (1983) for use in respect of technology can be extended to encompass science.

'Technology' may be divided into 'technology-as-practice' and 'technology-as-outcome'. Pacey (1983) has defined what he calls 'technology-practice' as the application of scientific and other knowledge to practical tasks which most commonly address the improvement of the physical conditions of human life. It consists of three, simultaneously operational, elements: the technical aspect, the organisational aspect, and the cultural aspect. The technical aspect includes the application of knowledge, skills and techniques, through the media of tools, machines, chemicals and living material, leading to the fabrication of products and, inevitably, to the generation of wastes. The organisational aspect is concerned with the social organization of industrial activity within economic frameworks, including that of the professionals undertaking the activity, the users and consumers of the products. These are the social organisations in which technology, as an activity, takes place, together with those that support, in one way or another, the conduct of that activity. The cultural aspect consists of the goals that are pursued, the values and ethical codes which underlie their adoption, the notions of social progress to which they relate, and of the conception of human creativity on which they rest. What emerges from Pacey's ideas is that technological process consists of thoughtful actions by individuals, taken within social contexts, to produce solutions to problems which they intend to be valued.

A similar terminology can be applied to science-as-process, which is concerned with finding explanations (science-as-outcome) for natural phenomena in the world as experienced. The technical aspect of science is the sum of the procedures whereby scientific inquiries are conducted. The organisational aspect includes the social institutions in which science is conducted (for example, universities) and through which the validity of the explanations produced is recognized (that is, scientific journals with their refereeing systems). The cultural aspect, often overlooked in studies of science, is concerned with the values that underlie decisions about the definition and selection of phenomena for study. It also concerns the explanations that are accepted by the community of scientists and presented as cultural knowledge. Science-as-outcome consists of a broadly conceived notion of 'scientific methodology' (the technical aspect). If scientific method is distinctive of science, it has been exported to other domains of human inquiry, for example, to the study of social phenomena. Science-as-outcome also consists of descriptions of how the material world behaves and of ideas about the entities of which the world is believed to consist or concepts by which it can be reliably analysed. Proposals as to how these entities are physically and temporally related to each other may be thought of as models and general sets of reasons for these behaviours, concepts, and models may be thought of as theories. Science thus consists of thoughtful actions by individuals within social contexts producing explanations of the natural world which they hope will be valued.

The similarity of these overviews of technology and science suggests that there are relationships between them (Gilbert, Boulter, & Elmer, 2000). It is possible to argue that the processes of technology first provide solutions to problems (outcomes), the success of which science explains later. For example, steel was initially developed by trial and error as a way of producing wear-resistant iron, whilst the effects on the structure of iron of the addition of small amounts of other elements such as nickel were only explained long afterwards. It is possible to argue that science precedes technology in time: technology is thus the application of science. For example, inquiries into the sequences of peptides within genetic material are leading to the rapid development of the biotechnology industry. A third interpretation is that the two are intertwined, as Barnes (1982) argues. Both involve invention, being creative, constructive, activities conducted within social contexts which draw extensively on prior achievements and which are subject to no one major constraint on their success. There is a traffic in knowledge and skills between the two, whilst they are both concerned to achieve definite outcomes. In short, science and technology are interdependent. One might expect this interdependence to be readily evident in the exhibits in a science and technology centre.

However, a walk around a science and technology centre suggests that interactive exhibits are often built around phenomena which illustrate particular concepts and models in science, most commonly drawn from physics or earth science. There is relatively little biology illustrated, perhaps because it is difficult to incorporate biological material into safe, robust, low maintenance, interactive exhibits. Whilst some elements of science-as-process are included, the treatment is usually far from complete. The relatively few exhibits of an apparently exclusively technological orientation are focused on technology-as-product but even here attention is often drawn to the underlying scientific concepts and models (again, mainly drawn from physics). There is plenty of evidence that the public pays attention to the implications of technology-as-outcome for matters close to their personal interests, drawn from everyday life, and to the science-as-outcome which underlies those examples (Lambert and Rose, 1998). However, a science and technology centre which sought to do justice to its potential remit might well pay more attention to science-as-process, to biology as a domain of inquiry, to technology-as-process, and to the relation between these three.

Doing so will entail a clarification of the purposes for which the science and technology centre exists.

1.3. Understanding, Literacy, or Awareness?

Some centres are evidently totally committed to the entertainment facet of their role, making no explicit attempt to educate. However, where they do so it is to develop either an understanding or a literacy in respect of science and technology.

The assumption that science and technology centres promote what is termed the public understanding of science has so far dominated discussion about their educational purposes (Thomas and Durant, 1987). The public has been shown to be deficient in an understanding (but, in reality, a simple knowledge) of facts and theories deemed to be the core of scientific achievement (Wynne, 1993). As a consequence, centres have been thought to be ineffectual in promoting such an understanding. This conclusion follows from the assumption that the 'public' has reason to learn these facts and theories in the form appropriate to the mode of assessment used to gauge understanding. There are no grounds for thinking that this assumption is justified. This purpose, then, is misconceived.

The notion of 'scientific literacy' seems a more realistic basis for viewing the purposes of science and technology centres, incorporating as it does ideas about the processes by which science is conducted and by which results are agreed as valid, as well as an understanding of science-as-outcome, (Millar, 1996). Alas, as mentioned above, these latter two ingredients are underplayed in or absent from most interactive exhibits, as often is much of the technology dimension. A 'scientific literacy' approach is, then, underdeveloped at present.

The germ of an achievable set of purposes is contained within Turney's (1996) view. Turney suggests that it is only when people can participate in the conduct of science and technology or, more realistically, in learning what science and technology can achieve, that they will begin to comprehend their nature in the 'scientific and technological literacy' sense. The suggestion that successful interactive exhibits are those which facilitate choice, access, and engagement, led Gilbert, Stocklmayer, & Garnett (1999) to see the issue as one of the design and presentation of interactive exhibits so as to promote the public awareness of science and technology:

> A person's awareness of science and technology is that set of attitudes and dispositions held concerning science and technology both as process and as outcome.

These attitudes and dispositions attach meanings to situations and tasks that the person encounters. They activate associated sets of actions and the skills with which the tasks are then carried out. Given the wide range of personal awareness of science and technology that is possible, very varied responses are also possible across a population of individuals when a particular situation is encountered. When a person holds a very positive scientific awareness, that situation will readily be seen to call for the use of scientific/technological knowledge and skills. At the opposite extreme, with a very negative scientific awareness, that possibility will not be seen at all. The extreme forms of action that will result will be engagement and avoidance respectively. Such a perspective does anticipate that the learning of science and technology *can* take place, but recognises that the development of more positive attitudes and dispositions is a co- or pre-requisite for personal progress. This development will take place in contexts to which the individual can attach positive personal meaning: the entertainment and educational dimensions have to be drawn together. If the ramifications of the public awareness of science and technology are to be explored, it must be done by using a perspective on learning which enables the ideas of content, processes, and attitudes to be integrated. The idea of 'model' has such a capability.

1.4. Models as a Basis for Entertainment and Education.

A model in science/technology is a representation of a phenomenon initially produced for a specific purpose. Gilbert, Boulter and Elmer (2000) give a broad discussion of the field from which this sub-section is largely drawn. The specific purpose for which any model is originally produced in science is to simplify the phenomenon so that the model may be used in inquiries to develop explanations. On the other hand, the specific purpose of a model in technology is the pursuit of solutions to problems. Models are severally derived from:

Entities that are material objects viewed as if they have a separate existence (e.g. of an aeroplane engine) or as if they are part of a system (e.g. an engine attached to an aeroplane). Such a model of an object can be either smaller than the phenomenon which it represents (e.g. of a railway train), or the same size (e.g. of a domestic pet), or bigger than the phenomenon (e.g. of a bacterium).
Abstractions, ideas, treated as if they are objects e.g. a magnetic field, or evolution. A model can consist both of entities that are material (e.g. an electrical conductor) and which are treated as if they are material (e.g. a magnetic field acting on that conductor).
A system, a series of entities in a fixed relation to each other (e.g. the components

and the connections between them in a diagram of an electric circuit).

An event, a time-limited segment of the behaviour of one or more entities in a system (e.g. a model of the movement of a component on a car production line).

A process, one or more events within a system which have a distinctive outcome (e.g. of the Bosch-Haber Process).

Models derived from all these sources have a place in science and technology exhibits in interactive centres. A classification of the ontological status of models is possible, only some parts of which are relevant here (see Gilbert, Boulter, & Elmer, 2000):

A *mental model* is private and personal cognitive representation. It is formed by an individual either alone or within a group. The representations formed by visitors using all exhibits in science and technology centres, interactive or not, are of the mental model variety.

An *expressed model* is placed in the public domain by an individual or group, usually for others to interact with, through the use of one or more modes of representation (see next paragraph). Any reflective person who has set out to express a mental model will be aware that the act of expression has an effect on that model. The act of expression both presents it to others and changes it. An individual visitor trying to explain an exhibit to him/herself, and especially the talk within a group of collaborating visitors, produces expressed models.

Different social groups, after discussion and experimentation, can come to an agreement that an expressed model is of value, thereby producing a *consensus model*. For example, scientists produce a wealth of expressed models of the phenomena that they are investigating. An expressed model that has gained acceptance by a community of scientists following formal experimental testing and subsequent development, as manifest by its publication in a refereed journal, becomes a *scientific model*. It then plays a central role in the conduct of scientific research for a length of time that is governed by its utility in producing predictions that are empirically supported. A *technological model* is a model of a proposed outcome of technology-as-process, the solution to a problem. Up to date exhibits are almost always derived from scientific or technological models.

Those consensus models produced in specific historical contexts and later superseded for many scientific or technology purposes are known as *historical models*. In reality, most of the exhibits in science and technology centres are derived from historical models.

That version of a consensus or historical, scientific or technological, model that is designed for educational and/or entertainment purposes, often after some further simplification, is a *curricular model*. Such models usually actually underpin the exhibits.

As the understanding of consensus, historical, and curricular models is often difficult, specific *teaching models* are developed to assist in that process. *Teaching models* are important in that they often form the substantial basis of the design of an interactive exhibit. Moreover, visitors produce them as a way of understanding interactive exhibits;

One or more of five *modes of representation* are used in expressed models (see Gilbert, Boulter, & Elmer, 2000):

The *material mode,* as the name suggests, consists of the use of materials e.g. a clay model of a kettle, a ball-and-stick model of a molecule. Almost all interactive exhibits are constructed in the concrete mode.

The *verbal mode* consists of the use of metaphors and analogies when talking about models. These are what visitors use when discussing the exhibits.

The *mathematical mode* consists of mathematical expressions, including equations e.g. the Kirchoff Equations. Only a few, the more 'advanced' exhibits, make use of the mathematical mode.

The *visual mode* makes use of metaphors and analogies in written form, of graphs, of diagrams (pictorial graphs), or indeed of several of these at the same time. These are usually found in the text accompanying exhibits.

The phrase *'symbolic model'* includes the visual, verbal, and mathematical modes.

The *gestural mode* consists of actions e.g. movements of the hand, of an object. The reaction of exhibits to the actions of visitors are often in the gestural mode e.g. using a 'distorting mirror', the spirograph.

These types of model and modes of representation are used to varying degrees when designing and using interactive exhibits,

2. SCIENCE AND TECHNOLOGY CENTRES AS PROVIDERS OF EXPLANATIONS

As has been argued above, science and technology centres are places in which scientific/technological processes and outcomes are explained to visitors in such a way as to engage their active involvement. We therefore need to have a way of talking about 'explanations', including some notion of what factors will determine their effectiveness. This is because, if the interactive exhibits are science-oriented, explanation will be a major function for them. If they are technology-oriented, it is likely that the associated science, with its attendant explanations, will be invoked. We also need to have a methodology through which to enquire into the effectiveness of the explanations that are put forward in any science and technology centre.

2.1. An Explanatory Framework

Gilbert et al (1998) suggested that 'an explanation is the answer provided in response to a specific question'. There are two parties to any explanation: the questioner and the explainer. The terms 'questioner' and 'explainer' may apply to either participant in the use of an interactive exhibit: the visitor (alone or with others) and the interactive exhibit itself (the task expectation and the explanatory text). The *appropriateness* of an explanation is an overall evaluation of it, arrived at by conjointly considering three components. The *suitability* of an explanation is a statement about the relationship between the type of explanation provided and the type of question posed. The *relevance* of an explanation is a measure of the extent to which it meets the needs of the questioner. The *quality* of the explanation provided is a measure of its scientific/technological standing. A judgment of appropriateness is initially made when the question is asked and the explanation is provided, but may be changed later in the light of further consideration of its value and significance. The appropriateness of the explanation provided by a particular interactive exhibit will be judged by each of the parties with an interest in it - the management staff of the science and technology centre, the exhibit designer(s), education staff and explainers, and the visitors.

If the suitability of an explanation rests on the use of the correct type of explanation, then a classification of explanations is required. A five category typology of explanations can be constructed, based on the nature of the questions which give rise to them (Gilbert et al 1998):

An *intentional* explanation is a response to the question 'why is this phenomenon being explained?' The explanation will include a statement of the purpose being addressed and give some idea of the importance of the phenomenon. It necessarily also carries some definition of the occurrence, the scope and boundaries, of the phenomenon. All too often, interactive exhibits do not explicitly either call for or give intentional explanations.

A *descriptive* explanation is the response to the question 'what are the properties of this phenomenon'. It always includes a naming of it. Most commonly, a descriptive explanation is a summary of the measurements (whether qualitative or quantitative) that are either made by the visitor or which are given in the text attached to the interactive exhibit.

An *interpretative* explanation answers the question 'of what is the phenomenon composed?' Such an explanation often assumes the actual existence of entities that are incapable of direct observation. Statements are made not only about their nature but also about how they are distributed in space and how that distribution changes with time. For example, an interpretative explanation of the allotropy of carbon - 'how is it that the element carbon has several forms'? - is that an allotrope consists of atoms of an element which exist in a particular geometrical arrangement, of which several are possible for carbon.

A *causal* explanation is a response to 'why does the phenomenon behave as it does?' A mechanism is proposed by means of which the phenomenon produces the observed behaviour through the operation of cause and effect on the entities of which it is composed. A mechanism can operate deterministically, in that one cause produces a single, definite, event: for example, the persistent application of sufficient force to an object will cause it to accelerate. Otherwise, it may operate probabilistically, in that one cause may produce many, possible events. For example, a change in the physical environment may cause the reproductive status of individuals within a species to alter. This is because there are a number of reproductive characteristics, randomly spread across the population, which are capable of responding to that environmental change.

A *predictive* explanation answers the question 'how will the phenomenon behave under other, specified, conditions?'. For example, the rate of a particular chemical reaction at some future moment in time can be successfully predicted because models of chemical kinetics have identified both the nature of the entities involved and the mechanisms by which change takes place. This might be thought of as a subset of descriptive explanation. It has been included, however, because predictions play an important part in the evaluation of the explanatory adequacy of theories and models.

3. TOWARDS A UNIFIED MODEL OF EDUCATION AND ENTERTAINMENT

Such a model, with which to provide a broad rationale for the personal awareness of science and technology in such centres, would address the following issues:

The range of opportunities presented to visitors. This would represent the factors that were borne in mind in making each particular interactive exhibit available for visitor use;

The basis of choice of a particular exhibit by a visitor. This would represent the factors that influence any one visitor's positive choice of one exhibit amongst the many with which it is usually surrounded;

The way that a visitor uses an exhibit. This would present the factors that influence what was done and for how long the visitor persisted with the exhibit;

The short term consequences for a visitor of using an exhibit. This would be an indication of a visitor's awareness of science and technology upon beginning to use the exhibit;

The longer term consequences for a visitor of using an exhibit. This would give an indication of the impact of the experience on the visitor's awareness of science and technology.

This Chapter cannot address all these factors in full. However, each of them can be discussed in brief with pointers given to future research work, whilst some are discussed more fully. The intention is to provide the groundwork for the subsequent evolution of the model.

4. THE RANGE OF OPPORTUNITIES PRESENTED TO A VISITOR

Six factors seem relevant here:

The design and manufacture capability of the science and technology centre. Ideally, all science and technology centres would have full design teams, in the sense of having of a capacity to develop ideas using a very wide range of media.

Equally, they would have the practical skills to realise designs for visitor use. In reality, only relatively few are sufficiently well funded to have both these capabilities: some will have one of them (often the skills of realisation, for plans developed elsewhere) whilst many will have none and will rely on the purchase of standard exhibits from other centres. The status of a centre in respect of these two capabilities will govern the operation of all the other factors under this heading.

The phenomenon being displayed. This can be framed separately by the concerns of the science, for example, the refraction of light by a prism; or by those of technology, such as the analysis of the electromagnetic spectrum from a distant star; or by those of everyday life, as in explaining the colours of the rainbow. The desirability of linking science and technology has already been discussed, whilst a direct reference to everyday life will provide a connection to common experience. This link may be made through a treatment of the applications of an idea in science to technology and through its implications for everyday life. In reality, these connections will be for many visitors on a range of 'many' to 'few' and on a range of 'familiar' to 'unfamiliar'.

The model of the phenomenon underlying the exhibit. In the terms discussed above, this may be a scientific or a technological model, although this would imply that the exhibit was up to date. It is much more likely that it will be based on an historical model. As all such models are often difficult to understand and equally difficult to incorporate into an exhibit, it is usual to use a teaching model based on an analogy to some aspect of common experience with which the visitor is likely to be familiar. The basis of this analogy may fall upon a range of 'near', in that it is very similar, for example, to.the use of a mist to represent the movement of air in a tornado, to 'far', in that it drawn from a different domain of experience, such as the use of a light wave to represent a sound wave.

The assumption made about visitors' prior knowledge. It does seem unlikely that the designer of an exhibit would assume that a visitor had a high level of scientific knowledge about the phenomenon on which it is based. Many exhibits assume no prior knowledge, although this would seem progressively less justified in the information-rich environments prevalent in many countries. It is less likely that the exhibit will have been designed in the light of the 'alternative conceptions' of scientific ideas that have been shown to be so prevalent in people of all ages (Gilbert and Watts, 1983).

The range of experimentation possible. All use of all interactive exhibits will, of their nature, require the visitor to take some action. However, the range will be from a heavily prescribed action to one where the visitor has considerable scope to choose between a range of options. In all cases the visitor observes what happens when the action takes place. The requirement to evaluate the consequences of the initial action and to decide upon and take subsequent action varies enormously.

The range of possible explanations of the phenomenon. The design of the initial activity proposed will govern the range of types of explanation that can be produced by the visitor. All exhibits will require descriptive explanations - that is, statements of what happens as the action takes place. There will, however, be a wide range of requirements for interpretative, causal, and predictive, explanations.

5. THE BASIS OF CHOICE AND USE OF AN EXHIBIT

Confronted by a wealth of exhibits in most science and technology centres, the visitor has to make a choice as to which to use, and in what order. The basis of decision-taking is undoubtedly complex, but seems to include the relation of an interactive exhibit to the visitor's existing understanding of science and technology, as well as the position and visual attractiveness of it. This will demonstrated by the interest taken in the interactive exhibit, the closeness of attention paid to the instructions provided, the self-confidence shown in following those instructions, the willingness to carry out follow-up actions in the light of the initial response to the interactive exhibit and the evident engagement of prior experience and knowledge.

6. THE SHORT-TERM CONSEQUENCES OF EXHIBIT USE

6.1 Enquiring into indivdual's awareness of science and technology

It has been argued above that an individual's awareness of science and technology will be made manifest in what they choose to do when confronted with a situation (here, an interactive exhibit) which engages attitudes and demands the use of skills. To form a view of that awareness requires that the individual has scope to explain personal views, decisions, and acts. The individual interview, with all its strengths and weaknesses, seems the most cost-effective way of getting at the thoughts and acts of individuals.

The remainder of this chapter draws on a pool of some 150 interviews conducted in Questacon, The Australian National Science and Technology Centre, in Canberra, over the period 1997-1999. The interviewees were all recreational adult visitors, for the usual meaning of 'public' refers to adults. Individuals or pairs were approached immediately after their use of an exhibit that seemed to have engaged their sustained attention. A more systematic treatment of the data obtained is given in Gilbert, Stocklmayer, & Garnett (1999).

6.2. The natures of some exhibits that engaged sustained attention

The nature of those exhibits, responses to which are quoted later, are summarized as follows.

Earthquake is a platform on which several visitors sit whilst it is rocked to and fro by a roller mechanism beneath it. It is intended that the visitor enjoys a simulated kinesthetic experience of what it would be like to be present on ground which is quaking. It is a teaching model in a material mode. The intentional explanation for it is provided in the associated graphics whilst it is designed to provide a descriptive explanation – that is, how it feels to be in an earthquake.

Tornado simulates the production and movement of a spiral of air moving circularly and transversely at high speed. The phenomenon is represented by a column of water vapour, produced by a generator, which is sucked upwards by an extractor fan at the top. The column is flanked by four vertical tubes, through which lateral air jets impart the circular motion. It is a teaching model in a material mode. Whilst the descriptive explanation is the most evident, interpretative and causal explanations are provided in the associated text.

Hand Battery demonstrates the production of electricity when two different metals are placed in an electrolyte and connected by an external circuit. A visitor's hands, sweaty to some extent and representing the electrolyte, are placed on sheets of different metals, with the body acting as the external circuit, the consequent microvoltage being shown on a dial. It is a teaching model in a material mode. The descriptive explanation is readily in evidence, whilst the associated text makes some attempt to provide a causal explanation.

Light Harp consists of a series of small holes, in each of which is embedded a photosensor. About 1m above the sensors are light emitters whose beams, shining onto the sensors, may be interrupted by the visitor's hands. When this occurs, electronically generated musical sounds are heard. The visitor may 'play' the light harp by moving her/his hands across the set of holes, a control enabling the set of sounds to be located within various 'sources' (e.g. 'strings', 'woodwind'). It is a teaching model in a material mode. The emphasis is on providing a descriptive explanation, although the range of use of the exhibit that is possible could readily lead visitors to produce predictive explanations.

Roller Race relates to the rotational inertia of circular objects when rolling down an incline, depending on the distribution of the mass about the axis of rotation. Three objects of the same mass (a ball, disc, ring) are simultaneously released down an inclined slope. The visitor is asked to predict the order in which they will reach the bottom of the incline, to do the experiment, and to explain the result. This is a teaching model in a material mode. Descriptive, causal, and predictive explanations are anticipated in its design and in the associated text.

Polarised Light invites visitors to place a polariser in front of a beam of light and to observe various plastic objects (e.g. forks, rulers, curved shapes). The visitor is invited to discover that various orientations of the polariser cause different patterns in the field of view; the consensus scientific model of plane polarized light underpins this exhibit. This is an example of the phenomenon, rather than a model of it. Descriptive explanations are the only ones that seem anticipated for a phenomenon with a complex causal explanation.

Black Hole consists of a metal bowl (representing a gravitational field) into which metal balls (representing any object) are dropped. The balls fall down the sides of the bowl, gaining speed as they do so, and 'vanish' down the hole at the bottom. This is a teaching model in a material mode. A very high level of inference is needed if the exhibit user is to progress beyond the self-evident descriptive explanation.

6.3. Visitors' immediate reaction to exhibit use

These are the outcomes, as revealed by interview, very shortly after the interactive exhibit has been used. They may be divided into four overlapping sub-classes. They are here presented in the order in which they may occur for the visitor: the *conative* outcomes, the quality shown in the planning for and engagement in subsequent related activites; the *behavioural* outcomes, the physical actions taken; the *cognitive* outcomes, the types of explanations produced and the nature of the conceptual understanding shown; and the *affective* outcomes, the feeling of access, of aesthetic appreciation, demonstrated.

6.3.1. Conative.
This type of response, immediately following the use of interactive exhibits, was rare. It may be that visitors do not perceive themselves to have the opportunity to plan an activity, perhaps because they view 'following instructions' as an automatic process or perhaps because the activities are often of short duration.

6.3.2. Behavioural
Again very few examples of this type of outcome were found:

> The children have come away with an interest in science, because they don't get an opportunity living in the country to experience these activities (Female, 30-39 years)

The reasons may be similar to those conjectured for conative outcomes.

6.3.3. Cognitive outcomes
There were no examples of intentional explanations, which was to be expected as the phenomena experienced were defined by the exhibits used. There were many examples of explanations of the descriptive type only. In the case of *Earthquake*, a quantitative description was provided

> It is useful, a good illustration of the Richter Scale, how it is measured, the fact that it doesn't just go up, but is more of an exponential. I didn't know that before (Male, 40-49 years)

but explanations were more usually qualitative, as would be expected from the nature of the exhibits used. For *Hand Battery*:

> ...if you put both on aluminium or both on copper, whether you'd get electricity, but you don't. You have to put one on copper and one on aluminium and it goes up a bit if you warm your hands up by breathing on them (Female, 30-39 years)

and *Tornado*:

> ...the other thing that we didn't realize that when you compare the US and Australia–they both have tornados (Female, 60-69 years)

and *Light Harp*:

> I just thought it was breaking the amount of light coming down onto the bar and from
> there I don't know how it works, how it changes tunes or anything. We enjoyed it.
> (Male, 60-69 years).

Some were descriptive, together with that most primitive of interpretations, the 'law'.
All 'laws' were erroneous. For *Roller Race*:

> Heavier things go quicker, it always happens (Male, in his teens)

A few interpretative explanations were imported as prior knowledge:

> It's good–there are different drag coefficients, wind drag, resistance, friction (Male, 30-
> 39 years).

although some were constructed on the spot. For *Hand Battery*:

> I didn't know that my hands were like salt water. I wondered if it was real electricity.
> It seems to be the effect of my hands that produced the effect (Male, 30-39 years)

Causal explanations did seem to be obtained, whether generally:

> I'll take away a lot more knowledge, I suppose, of things. You see how things work..
> (Female, 30-39 years)

or in respect to specific exhibits. Thus for *Roller Race*:

> The ball rolls faster because it can use its energy for rolling fast - the energy is used
> more for speed than rolling. The disc came second because it needed a bit more energy
> for rolling and the little cylinder came last because it obviously needed most energy to
> roll and couldn't keep up the speed as well! It teaches you the way that different objects
> use the same energy for rolling and some for speed, depending how the mass is divided
> up in the different objects (Female, 30-39 years, graphic designer).

for *Light Harp*:

> I found out that it's operated by the light going through the beams and as soon as the
> light shut down, covered up, it sent signals to the computer and it starts up, starts
> different music.

and *Tornado*:

> I was wondering why it spins. I assumed it was the Earth's gravity that was making it
> spin. The turning of the Earth. But I'm not sure that's why. Whether it's just a natural
> thing when you get the warm air and it goes up... I'd like to know if it's different in the
> Northern Hemisphere, whether it spins the other way, and I'd be curious to see if you
> could replicate it spinning the other way, like put it in an opposite rotated environment...
> it's difficult to measure that, because we've tried to do it in the sink, when theoretically
> it should go the other way in the sink, and it doesn't always. (Male. 30-39 years)

There were no examples of genuine predictive explanations being put forward,
that is, proposing what behaviour would be in particular circumstances. However,
individuals did undertake speculative activity, to see what 'would happen if...' It

may be that they had some implicit idea of what would happen. For example, with *Polarized Light:*

> The depth of it, that was what I was looking at. Trying to work out how the depth and shallowness worked. I was interested in the colours, how blue colours and cool colours recede and warm colours make things approach you...I was trying to work out if that was happening on that (Male, 60-69 yrs)

Multiple explanations were put forward, but these may have arisen from prior knowledge. For example, an interpretative plus causal explanation for *Polarized Light:*

> I was also, as one often does when playing with diffraction stuff, turning it until it didn't let any light through. I was explaining to my wife that because polarized light is in a single plane, once you put them together at right angles to one another it lets no light through at all. (Male, 50-59 years)

6.3.4. Affective

This class of response evidently relates closely to the 'entertainment' aspect of the role of the science and technology centre. At the same time, Tyson, Venville, Harrison, & Treagust (1997) have shown the importance of 'the affective' in providing the dynamic for cognitive change. This class is thus important in the 'educative' element of the role of the centre.

There were examples of positive feelings arising from the effects of a physical action. For example, in respect of *Light Harp:*

> I enjoyed the exhibit, trying to place your fingers with the different parts associated with the piano and the organ. It wasn't just one note, one key, I tried all the different instruments (Female, 20-29)

in *Tornado:*

> It's unpredictable and it's beautiful to watch because it's so sinuous (Male, 30-39yrs)

and

> Int. What do think of this exhibit?
> Vis. I think it's absolutely fascinating
> Int. What were you thinking while you were looking at it?
> Vis. I thought what a clever thing, what a feat of engineering. To explain something that occurs in nature, I thought it was very apt. I had never thought about it before. (Male, 20-29 yrs)

Perhaps the most sophisticated responses arose where cognitive understanding was augmented by the provision of a psychomotor response. For *Earthquake:*

> What have I found out? The severity. When you try to relate to the quakes that they are mapping, you have to try and work out the severity on the Richter Scale. It gives you a bit of a feel for when you hear that an earthquake is 3, or 5, and you think–well, OK, I've sat through one that's about 2, so you get an idea of the severity of what people have to survive through... (Male, 30-39 years)

A feeling that the exhibits were simplifications was a reminder of the inherent limitations of the genre:

I think I would prefer something that is a bit more enclosed, like with walls, or even a house situation to feel what it would be like in a building, and perhaps have some props and things that were falling as it was shaking–here you are looking around and you're getting a feeling but you're not getting a complete feeling   (Female, 30-39 years)

For at least one person, the general feeling was one of approval, when compared with school:

It is good to let children see how simple and practical science can be. At school you learn formulas from books and see some experiments, but that way is less pleasant (than this) (Male, 30-39 years)

There was an additional class of initial outcome that can be termed 'workmanship'. It was an appreciation of, an affective response to, the design and construction of the exhibits as objects. One example of this class was:

Int. Could anything have improved the exhibit for you?
Vis. No, I think it's wonderful, very well done. Being in the business myself (as a technical illustrator), I appreciate the workmanship that goes in there, it's a top rate display, photography and illustrations, both for children and adults
(Male, 60-69 yrs)

7. THE LONGER TERM CONSEQUENCES OF EXHIBIT USE

A 15% sample of those interviewed immediately after their use of an exhibit were telephoned 6 - 10 weeks after the visit to Questacon.

In the light of the short-term outcomes it is not surprising that the longer-term outcomes were also confined to the cognitive and affective domains. What was perhaps more surprising was that, once liberated from the researcher's questions about 'what did it mean to you…' visitors were, in the longer term, inclined to give a wider range of response.

7.1. Cognitive

It did seem that some prior descriptive explanation of a phenomenon was useful when it came to consolidating the gains derived from a visit. For example, a visitor who had used *Black Hole* said:

I knew what a black hole was before the visit - I used to watch a lot of science programs in the U.K. This British astronomer…used to talk about black holes. I've been intrigued by black holes because we don't know what happens when you get into one (Male, 60-69 yrs)

An evaluation of the descriptive explanations of the experience could lead to a critical stance over at least some of the exhibits e.g. in respect of *Earthquake:*

I've been thinking about some of the things that I saw there, about Earthquake - the news about the earthquake made me think about it. I talked about it with my son and we talked about differences between the machine and Questacon and real life…in real life buildings, roads, everything would be crashing down around you. (Male, 30-39 yrs) and broadening of perspective in the exhibits was called for:

> ...I'm more interested in politics. If you can bring science into politics, then I'd be more interested in science. (Male, 30-39 yrs)

However, many found the descriptive explanation that they had met in Questacon adequate for their later experiences

> I've come across the Hand Battery before, the idea of it, so I showed my wife and explained about the continuity of the current through different metals. I haven't thought about Hand Battery since our visit, but I thought about it a lot whilst I was there - very interesting that you get a reading of electricity from your body (Male, 60-69 yrs)

particularly when that explanation could be rehearsed:

> I've thought about the Light Harp when I hear organ or harp music. I hear it and see what I saw when I was using it...experience it directly again. I don't see myself at it, I see myself doing it again (Female, 60-69 yrs)

In at least one visitor interviewed by phone, the need for a causal explanation had been aroused:

> I don't understand it (Roller Race)--it puzzles me and I keep trying to work it out (Male, 30-39 yrs)

One interesting reaction was to use the explanation that had been obtained within some technological application, whether original:

> the Roller race got me thinking about...the big wheel that once you get it up to speed, you don't have to put much energy in to keep it going. I was wondering if you made a solid ball in the engine, if you had a ball rotating in the engine, it would be more efficient. It makes you think about how things happen in real life (Male, 30-39 yrs)

or familiar:

> I haven't thought about (the Light Harp) - I knew the operation of it, so it wasn't new to me, it's just an alarm system. I'm an electrician. (Male, 20-29 yrs)

Another approach to using an explanation was to link it to another experience, usually obtained through a different medium:

> Yes, I've thought about it. Something I saw or read, or it might have been on the news, they were talking about the asteroid that might come to Earth...I thought, why don't they send it to a Black Hole (Male, 60-69 yrs)

Inevitably, the experience or the descriptive explanation of it was used to augment that provided by school. It is interesting that all the exhibits discussed with children were outside the chosen remit of this inquiry: perhaps the research had selected exhibits with 'adult' as opposed to 'child' interest. Thus, for a dinosaur exhibit, a woman (30-39 yrs) commented

> My daughter had a dinosaur project at school and it was helpful for that...The children loved the one where they tried to outrun the dinosaur, they ran as fast as they could, but they were surprised that, no matter how fast they ran, they would be eaten by it.

And, discussing the Tesla Coil

> The kids kept mentioning Lightning...they know what it looks like, have a picture in
> their minds, could see it

7.2. Affective

Positive affective responses had been engendered which had persisted over time:

> The Light Harp just came into my mind–I thought about playing (it) and the Explainer
> who clapped, gave me applause when I finished playing it. I just remember feelings,
> feelings of warmth, pleasure, enjoyment (Female, 60-69 yrs)

with evidence that access to science and technology had been facilitated:

> (following my visit to   Questacon), scientific phenomena are more interesting, less of a
> mystery, more understandable in terms of basic principles (Male, 30-39 yrs)

8. DISCUSSION

The value of any model is manifest in the avenues of inquiry to which it draws attention.   Some will be obvious, coming directly from a consideration of the elements of the model, some will be pragmatic, leading to an improvement in practice in the field of endeavour which is being represented. Those avenues that point to new areas of research, or even to a changed emphasis between existing areas of research, will be much valued. Last, and most important, the model should suggest the avenues by which it can itself be tested.

The discussion above of the elements that might contribute to a unified model of entertainment and education in science and technology centres does suggest a number of avenues for pragmatic improvement in respect of the design and use of interactive exhibits.   Inevitably, they might be associated with, or better still, preceded by, further research. One such area concerns the way that decisions are taken about the design, construction, and deployment, of exhibits. What value systems, relating to the balance of 'education' and 'entertainment' issues, are held by the varied members of the teams that design interactive exhibits? In what way does the capability of a science and technology centre to realise designs constrain the types of design that are agreed upon? How can these factors be optimised so that the range of exhibit types is maximised? Any success in this direction should improve the balance between educational and entertainment aims that are successfully addressed.

Similarly, the demonstration that there is a longer-term educational benefit to the use of interactive exhibits suggests that science and technology centres might make more effort to provide 'follow-up' activities. What can be incorporated into individual interactive exhibits, or perhaps within the overall provision made, that encourages visitors subsequently to rehearse their understandings to other people, to forge links with other (perhaps subsequent) experiences, to seek the analogical transfer of ideas to different contexts?. A combination of questions, suggestions, and challenges might do the trick.

The discussion of the relationship between science and technology given earlier does suggest a review of what is included in interactive exhibits. Can the 'institutional' and/or 'cultural' aspects of both science and technology be incorporated into interactive exhibits? If this could be done, the current emphasis, perhaps over-emphasis, on the technical aspects of both could be reduced, thus giving a more balanced interpretation of these fields of human endeavour. How can emphasis placed on science be properly reduced in favour of technology, a change which would enable experiences to be more directly related to the everyday experiences of visitors?

The area of models and explanations is ripe for more research and subsequent development. From what range of sources are the analogies drawn from which teaching models are constructed currently drawn? How can that range be extended, thus making the interactive exhibits accessible to a wider cultural background of visitors? In particular are there teaching models which are particularly effective in linking scientific explanations, technological outcomes, and everyday experience? How can visitors be best introduced to the 'codes of interpretation' of teaching models so that they may most fully understand the underlying consensus/historical models? Is full use made within interactive exhibits of the range of 'modes of representation' (concrete, verbal, visual etc) and are the modes employed orchestrated to best effect for the support of learning?

What can be done to improve the 'explanatory effectiveness' of interactive exhibits and their accompanying texts? How can the range of explanations focused upon be improved, for example, to include more 'intentional' explanations, perhaps focusing on future developments in science and technology? How can the appropriateness of explanations provided be improved? What can be done to increase the suitability of the explanations provided in respect of the range of questions actually asked by visitors? How can the relevance of those explanations to visitor's own questions be improved? How can the quality of the explanations provided be improved?

Karl Popper's great contribution to science was to suggest that no theory (and hence, no model) can ever be proved correct. The best theories (models) are those that most readily lend themselves to the posing of questions aimed at refuting them and which survive those attempts at refutation. Such a fate awaits the model that might be constructed from the elements suggested here. But perhaps not immediately! Let us first attempt to build it and to find confirmatory evidence for it. It should reflect all the facets of decision-taking, construction, use, and significance for individuals, of interactive exhibits. It should facilitate an exploration of all the facets which are suggested above as being important. It should enable the miscellany of existing relevant research to be drawn to together and evaluated. A long-term program of research and development lies in prospect.

School of Education, The University of Reading, UK

9. REFERENCES

Barnes, B. (1982). The science-technology relationship: a model and a query. *Social Studies in Science* (12) 166-172.

Gilbert, J.K., Boulter, C.,& Elmer, R. (2000). Positioning models in science education and in design and technology education.   In J.K.Gilbert and C.J.Boulter (Eds.), *Developing Models in Science Education*. Dordrecht: Kluwer, 3-19.

Gilbert,J.K.,Boulter,C.J. & Rutherford,M .(1998). Models in Explanations, Part 1: Horses for Courses. *International Journal of Science Education*, 20 (1), 83-97.

Gilbert, J.K., Stocklmayer, S. & Garnett, R. (1999). Mental modelling in science and technology centres: what are visitors really doing? In S. Stocklmayer & T. Hardy (Eds), *Learning Science in Informal Contexts*. Canberra, Australia: Questacon, 7-15.

Gilbert, J.K. & Watts, D.M. (1983). Concepts, misconceptions, and alternative conceptions: changing perspectives in science education. *Studies in Science Education, (10)* 61-98.

Kelly, G.(1963). *A Theory of Personality: the Psychology of Personal Constructs*. New York: Norton.

Lambert, H. & Rose, H. (1998). Disembodied knowledge?: Making sense of medical science, in A. Irwin and B. Wynne (Eds.), *Misunderstanding Science?: The Public Reconstruction of Science and Technology.* Cambridge: Cambridge University Press, 65-83.

Lave, J. (1988). *Cognition in Practice: Mind, Mathematics and Culture in Everyday Life.* New York: Cambridge University Press

McManus, P.(1992). Topics in museums and science education. *Studies in Science Education, (20)* 157-182.

Millar, R.(1996). Towards a science curriculum for public understanding of science. *School Science Review, 77* (280), 7-18.

Pacey, A. (1983). *The Culture of Technology*. Oxford: Blackwell.

Shortland, M.(1987). No business like show business. *Nature (Lond), (328)* 213-214.

Stevenson, J. (1991). The long-term impact of interactive exhibits.*International Journal of Science Education, (13)* 521-531.

Thomas, G.(1994). The age of interaction. *Museums Journal, 94* (5),33-34.

Thomas,G., & Durant,J. (1987). Why should we promote the public understanding of science. *Science Literacy Papers, (1)* 1-14.

Turney, J. (1996). Public understanding of Science. *Lancet*, (347) 1087-1090.

Tyson, L., Venville, G., Harrison, A., Treagust, D. (1997). A multidimensional framework for integrating conceptual change events in the classroom. *Science Education, 81* (4), 387-404.

Wynne, B. (1993). Public uptake of science: a case for insititutional reflexivity. *Public Understanding of Science, 2, (4),* 321-337.

9. REACHING THE PUBLIC – COMMUNICATING THE VISION

On the 23rd February, 2000, the British House of Lords published the findings of a Select Committee on Science and Society. The report begins by stating that 'society's relationship with science is in a critical phase', citing recent developments in biotechnology and the BSE (mad cow disease) disaster as eroding public confidence and creating public unease.

The list of witnesses to this Committee reads like a *'Who's Who'* of science communicators in Great Britain. Several are contributors to this book. The overall findings and recommendations of the Committee are an ideal recipe for science communicators to build bridges between science and the public, including recommendations for serious funding to support the activities of interested bodies such as the Committee on the Public Understanding of Science (COPUS) and the British Association (BA).

It is an extraordinary document, given that the broad base of practicing scientists still does not hold communication in high esteem, except with their peers. Early in 2000, the recipient of the Young Australian Scientist of the Year Award, Dr Kirsten Benkendorff, stated to a reporter of *The Australian* newspaper that

> some scientists had a 'cynical attitude' that a researcher who took time off from usual work to communicate his or her findings to the media was not being a good scientist...

She said that one academic had told her that awards of this kind were a recognition of her success as a communicator rather than of her science. She added that she was proud of her success and remarked that it was the first time she had realised it could be detrimental to a career in science (Patrick Lawnham, pers.comm.).

The findings of the Select Committee include recommendations that all scientists have training in communication and understand the social context of their research. Since 1996, the team at the Australian National Centre for the Public Awareness of Science has been running workshops for research scientists in communicating their discipline to all sectors of the public and to their peers. It can be an uphill task, in that the attitude expressed to Dr Benkendorff is evident in many of their clients. Once convinced of the need for change, however, the scientists become committed to the task of 'translating' their research into interesting, relevant, contextual information. At the end of the three day workshop, no one who has attended (and the number now exceeds 200) has been found to be researching something too intrinsically boring or too difficult for the ordinary person to understand. And all the scientists are passionate about their research.

S.M. Stocklmayer et al. (eds.), Science Communication in Theory and Practice, 143–148.

Part of the workshop deals with the overall aims of reaching the public. Discussions about the traditional social, practical and cultural demands for a scientifically literate public follow the lines of the justifications of the Royal Society quoted in Irwin and Wynne (1996, p.5). Improved understanding, it is claimed by the Royal Society, will benefit:

> *national prosperity* (for example, a better trained workforce)
> *economic performance* (for example, beneficial effect on innovation)
> *public policy* (informing public decisions)
> *personal decisions* (for example, over tobacco, diet or vaccination)
> *everyday life* (for example, understanding what goes on around us)
> *risk and uncertainty* (concerning nuclear power or BSE)
> *contemporary thought and culture* (science as a rich area of human inquiry and discovery).

Like Irwin and Wynne (pp.5-10), however, the team stresses the diversity of audiences, the variety of understandings and knowledge, the variety of interest and engagement and the impossibility of one-way, conduit-style transmission. At the same time, though, they tell the scientists about the many surveys which have been conducted worldwide, including in Australia, which state that the public likes to read about science and is proud of it. Such surveys are quoted also in a British Select Committee's Report: 35% of the British public, for example, state that they are very interested in new scientific discoveries. Only 29% are very interested in sport. Similar figures are quoted in other parts of the world. Yet the Select Committee finds wide mistrust of scientists, especially Government ones, and that this is 'breeding a climate of deep anxiety among scientists themselves' (p.1).

Where, then, are we going wrong? The scientists in the workshops begin by blaming the schools: 'If the subject were only taught better... we wouldn't have to worry about the public'. It is with some difficulty that they move to the position of accepting that teachers are scientists too, and that the curriculum is decided very often by tertiary demands. There is to be no easy way out of the dilemma, and no culprits.

We believe that, for the public, the issue is one of ownership, complicated by problems of access. There is little doubt that, when the occasion demands, ordinary 'unscientific' people can become very expert, very knowledgeable, very much in command of the science. With respect to public understanding, there have been many authors who have examined in detail case studies of public crises in which those concerned rapidly (albeit with some pain) became 'experts' on the issue at hand. In their book *Misunderstanding Science? The public reconstruction of science and technology*, Irwin and Wynne (1996) recount a number of instances in which this has occurred. For example, the hill sheep-farmers in the Lake District, when faced with radioactive contamination following Chernobyl, 'entered the scientific arena... redrew its boundaries, and, operating with different presuppositions and inference rules, also redrew its logical structures' (p.31).

The problem, we believe, in using content knowledge as the criterion of public literacy is that it, by definition, produces the 'deficit model' described by Wynne (1993). The Select Committee of the House of Lords shares this concern. A recurring theme in the Report is the value of the phrase 'public understanding', which the Report interprets to mean 'the understanding of scientific matters by non-experts'. Bryant (1998) has defined it in more detail:

> The public understanding of science is the comprehension of scientific facts, ideas and policies, combined with a knowledge of the impact such facts, ideas and policies have on the personal, social and economic well-being of the community.

The connotations of knowledge and comprehension of facts have been explicitly stated by the Select Committee as problematic:

> It is argued that the words imply a condescending assumption that any difficulties in the relationship between science and society are due entirely to ignorance and misunderstanding on the part of the public; and that, with enough public-understanding activity, the public can be brought to greater knowledge, whereupon all will be well. This approach[27] is felt by many of our witnesses to be inadequate; the British Council went so far as to call it 'outmoded and potentially disastrous'. (p 140)

The Report goes on to urge a new term to replace this 'backward looking vision', perhaps in terms of 'science and society' because 'it implies dialogue in a way that public understanding of science does not'.

The recommendation that a new title for COPUS should reflect a dialogue with the public is underlined by all the authors of this book. We are left, however, with the task of finding an alternative term for 'Public Understanding' which reflects the need to communicate science more effectively, to share with the public the concerns and the issues and to convey a sense of individual change. We therefore propose the term 'public awareness of science and technology' to cater for this individuality and to remove connotations of concrete measurement - and of a deficit model. We have defined this term as follows (Gilbert, Stocklmayer and Garnett, 1999):

> Public awareness of science and technology may be defined as a set of attitudes, a predisposition towards science and technology, which are based on beliefs and feelings and which are manifest in a series of skills and behavioural intentions.
>
> The skills of accessing scientific and technological knowledge and a sense of ownership of that knowledge will impart a confidence to explore its ramifications. This will lead, *at some time*, to an understanding of key ideas/ products and how they came about, to an evaluation of the status of scientific and technological knowledge and its significance for personal, social, and economic life.

This definition therefore provides for personal development through many levels of understanding and makes no attempt to define what these levels, in terms of scientific content, might be.

The term was coined originally to help to define research in science centres which sought to probe visitor experiences beyond simple behavioural observations, but it seems applicable to most of the communication techniques described in this book. It carries with it an expectation that awareness will be enhanced through ownership and access, through personal experience and exploration, rather than by didactic scientific transmission of the traditional kind. It implies that no 'progress' can be made except through personal meaning-making, which is consistent with constructivist theory. It provides also for cultural differences within 'the public'.

In Australia today, increasing emphasis is being placed on the need for involvement of the community, not only in sharing the outcomes of research but in its design. This applies mainly, in 2000, to the agricultural and environmental areas of research, but the trend is increasing rapidly. Some funding agencies (for example, the Grains and the Wool Boards in Australia) are moving to a 'purchaser-provider model' in which those stakeholders, especially landholders, who are beneficiaries of the research also advise what kinds of projects should be addressed. There are many difficulties with this model, (Keen and Stocklmayer, 1999) but the practice is bound to grow as scientists, more and more, have to account for their activities to a discriminating public.

The face of science communication around the world will change greatly in the next few years. Increasingly, it is being recognised as an important element in the conduct of all scientific research. Research into science communication itself, however, also is required - to understand different ways of affecting the public awareness of science and to understand how the public reaches decisions on important issues. Such research is vastly underfunded worldwide. It is extraordinary that, on the one hand, large sums from the public purse are spent on analysing what is wrong, through surveys, committees (select or otherwise) and the like, yet serious research into the nature of public awareness and understanding is regarded as a peripheral, unimportant aspect unworthy of major funding. Such research is not classified as 'science' and it is not 'education' - so it is neither understood nor supported. Despite recognition of the many ways in which festivals, science centres, public organisations such as BA and COPUS, the media, the internet and so on are affecting public awareness in a positive and enjoyable way, the task of defining what they are doing and why they 'work' - or even whether they 'work' - is not addressed. This, too must change if we are to come any nearer to knowing how to reach the public, how to provide for that ownership and access we deem to be the goal, and how to make science everybody's concern.

A theoretical model for the public awareness of science has been suggested by Stocklmayer and Gilbert (in press). This model proposes an iterative process of learning, centred around successive encounters with the 'target' topic as shown in Figure 1. We have coined the term 'personal awareness of science and technology' (PAST) to allow for individual differences.

In this model, an individual (let us say a visitor) in a state of PAST 1, encounters an experience which enhances this PAST in some way. Links to this experience are, therefore, quite strong, and are informed by remindings and by the interaction itself. The result is a new level of awareness, PAST 2. The connection with the desired target may, however, be quite tenuous at the time of the interaction. A further experience related to the same target will be informed by a new, stronger set of remindings which will, in their turn, increase the connection to the target. This iterative process can occur indefinitely.

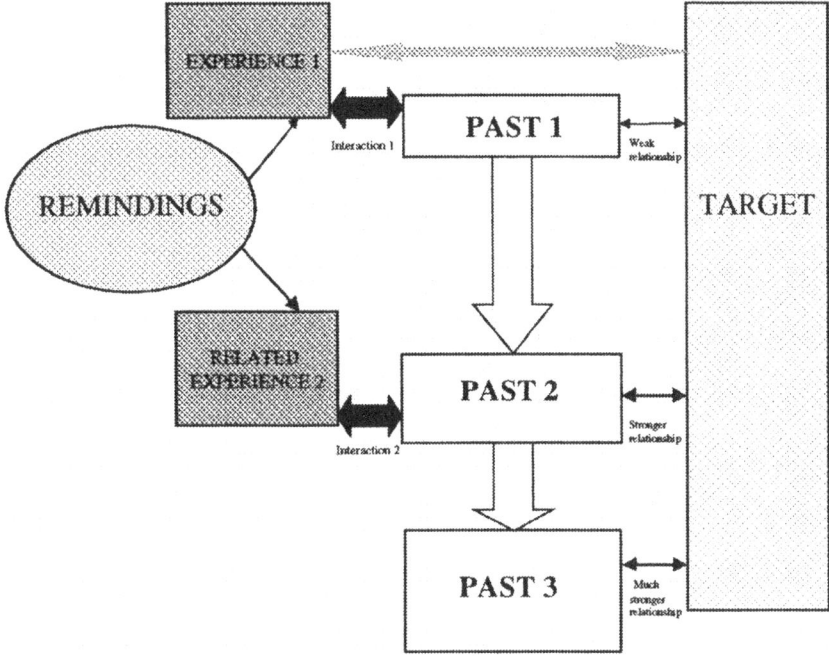

Figure 1: A model for the personal awareness of science and technology
(Stocklmayer and Gilbert, in press)

The model places all the emphasis upon the *experiences*, since they are the way in which the visitor encounters the underlying science. Good design of such experiences is critical – and good design implies facilitating interaction and understanding (Stocklmayer and Gilbert, in press).

The implication of the model is that effective increase in a person's awareness of science is achieved only by engagement with analogous experiences, whether they be interactive exhibits, a journal article, a festival, a television or radio program - indeed any kind of encounter with science in an accessible form. The challenge of science communication is to provide those contextual experiences, to provide for the link with the 'target' to be clearly delineated and to encourage further interaction. It is not a simple process - but to hope that the public will be motivated to 'learn' and remember without accompanying engagement - even entertainment - is surely to fail in the endeavour to which this book is dedicated.

National Centre for the Public Awareness of Science, Australian National University, Canberra, ACT 0200, Australia.

REFERENCES

Bryant, C. (1998). *A taxonomy of scientific communication.* (An address to the Management Committee of the Federation of Australian Science and Technological Societies).

Gilbert, J.K., Stocklmayer, S.M. & Garnett, R. (1999). Mental modeling in science and technology centres: What are visitors really doing? In S.Stocklmayer and T.Hardy (Eds). *Proceedings of the international conference on learning science in informal contexts.* Canberra: Questacon, 16-32.

House of Lords (2000*). Report of the Select Committee on Science and Society.* London: House of Lords.

Irwin, A, & Wynne, B. (1996) *Misunderstanding Science? The public reconstruction of science and technology.* Cambridge: Cambridge University Press.

Keen, M., & Stocklmayer, S. (1999). Science communication: the evolving role of rural industry research and development corporations. *Australian Journal of Environmental Management, 6,* 196-206

Stocklmayer,S., & Gilbert,J. (in press) New experiences and old knowledge: towards a model for the public awareness of science, *International Journal of Science Education.*

Wynne, B. (1993). Public uptake of science: a case for insititutional reflexivity. *Public Understanding of Science, 2,* (4), 321-337.

SECTION 3.

SCIENCE COMMUNICATION AND THE MEDIA

P. SPINKS

10. SCIENCE JOURNALISM: THE INSIDE STORY

'True descendants of Prometheus, science writers take the fire from the scientific
Olympus, the laboratories and the universities, and bring it down to the people.
William Laurence, former science writer, *New York Times.*
In S. Weart, (Ed.) (1989) *Nuclear Fear.* Cambridge: Harvard University Press.

Why should anyone wish to be a science journalist? On the face of it, those writing about the latest scientific developments for the print media enjoy one of the most rewarding and satisfying careers imaginable. The work potentially provides access to some of the world's keenest minds, involves probing pioneering discoveries about Nature and facilitates a kind of ongoing education by personal interaction. What is more, for those loving the limelight, reader surveys intimate that science stories are among the most popular in newspapers.

Science journalism may also be a professional passport to the kinds of travel and adventure usually reserved for a privileged few. In the course of two decades of broadcasting and writing about science, I have had the opportunity to go places and do things about which many people dream. They include flying a North Sea sortie at almost twice the speed of sound in a Lightning interceptor jet of Britain's Royal Air Force and touring key research facilities in Taiwan and Japan. In the United States, I visited NASA Mission Control in Houston and witnessed the latest generation of Boeing 777s being built in Seattle. I've also flown low across an icy stretch of Antarctica, scuba dived in the Strait of Hormuz, been treated to a geologist's account of the volcanic splendours of Samoa and straddled the International Date Line in Fiji.

In Australia, I've relished a botanist's personal helicopter tour of fire-scarred Kakadu National Park, explored parts of the rugged Red Centre and visited some of the southern hemisphere's most powerful telescopes in northern New South Wales. Into the bargain, I have received first-rate popular-science books to review and have met some engaging and exceptionally smart scientists, several of whom have become friends. That's been the good part.

Like the curate's egg, however, there's another side to science journalism that may be anything but paradisiacal. No truer is this than for those at the sharp end - the newspaper science writers, whose role I shall primarily discuss here. As many a journalist can testify, the daily routine of covering science for a newspaper is sometimes fraught with such formidable hurdles that even the most determined and dedicated among us may end up in public relations or as gardening correspondents - or, in dire cases, as gardeners.

S.M. Stocklmayer et al. (eds.), Science Communication in Theory and Practice, 151–168.

1. RESPONSIBILITIES

The newspaper science reporter, also called science correspondent, is responsible to the news desk, which is usually governed by a news editor and a bevy of acolytes. News minders sometimes assign selected stories to science journalists, who are also required to generate their own ideas and story leads. The correspondent's brief, essentially, is to relate the latest, most fascinating and important scientific developments to readers, for many of whom newspapers are the prime, if not sole, source of such information. Science - the biggest, most complex and, in my opinion, prestigious round on which to report, and also the least understood - spans anything from archaeology, botany and zoology to mathematics, chemistry and biotechnology, and much in between. It also encompasses science policy. The range is so vast as to deter most other 'rounds reporters', or 'beat reporters', as specialist journalists are termed. We all have our predilections, mine being physics, astronomy and cosmology. I have also gravitated towards developments in the earth sciences.

Although basic research often spawns medical or technological spin - offs, the science portfolio should not, in my opinion, include the worlds of health and general technology. Many science journalists, however, write frequently about health - related topics and consumer knick - knacks that provide likelier fodder for page one - even though health and technology may already be covered by one or more rounds reporters. This is because editors believe, rightly or wrongly, that people prefer reading about their bodies and state-of-the-art technological trinkets to more fundamental subjects such as the origin of the universe, the basic structure of matter, psycholinguistics or perhaps evolutionary theory.

In part, editors are swayed by surveys implying that readers find non-medical stories less gripping. It is uncertain how these studies square with those proposing that science articles rank among the newspaper favourites. The latter seem to be buttressed by the International Social Survey Programme's 1995-1996 'National Identity' poll, based on 30,894 interviews in 24 countries. It suggests that Australians are prouder of their nation's achievements in science and technology than most other nationals, except Americans. Only accomplishments in sport were rated slightly higher than those in science. Perhaps the importance attributed to the biomedical sciences relates to their being relatively well subsidised by government, industry, trusts and charities, while their somewhat impecunious cousins, the basic physical sciences, rely more on government funds alone.

As well as health and technology, some science reporters regularly cover aspects of the environment, such as endangered species, especially of the cuddly, furry variety. I have less difficulty with this and have reported on numerous ecological issues that are sadly neglected nowadays. Some so-called science journalists, however, wax lyrical about plants and animals at the exclusion of almost all else, probably because they make nice photographs, are relatively easy to write about and are readily understood by editors.

Besides covering the day-to-day world of science, correspondents may be required to provide in-depth features on scientific developments or trends. Occasionally, science news analyses or commentaries may be requested to boot.

Furthermore, science scribes are enjoined to provide 'explainers' when necessary. These are background accounts of any topic currently hitting the headlines which, in an editor's opinion, is somehow related to science, technology or the environment. Over the years, I've been requested to elucidate such natural phenomena as solar and lunar eclipses, meteorite showers, lightning strikes, sunsets, floods, droughts, El Niño, La Niña, the 'presence of sharks near the shoreline', cyclones and unusually hot, humid or cold days, weeks, months or seasons. Among unnatural phenomena, I've been asked to describe the workings of assorted missiles and explosive devices and, once, to draw on contacts to explain why a TWA jumbo jet exploded shortly after leaving New York for Paris on 17 July 1996.

Regular pages or sections devoted to science are quite rare these days, particularly in countries like Australia. This is partly because science, unlike information technology or medicine, is not an obvious source of advertising revenue. When such sections are launched, it invariably falls to science reporters to edit and sometimes even write the copy. This may be a full-time job in its own right, yet is still to be fitted in alongside covering news, writing features and the odd explainer or news analysis.

2. GREAT EXPECTATIONS

Science journalists are expected to be pretty much across the various disciplines within their vast domain. A sound grasp of the fundamentals and history of each subject is more or less taken for granted, as is a familiarity with key developments reported in major journals, not to mention some of the minor ones. Some demanding editors even consider that the conscientious science reporter should be aware of developments yet to be announced publicly. The (high) hope here is that the media may get to scoop the scientific press - even though respected international publications, such as the British journal *Nature* and its American counterpart *Science*, enforce strict embargoes (which grant science writers time to prepare articles on highly technical subjects). Editors certainly expect much, yet they infrequently return the compliment: science writers may not be consulted about running science stories and it is not unknown for their advice to go unheeded.

Scientists, too, harbour ambivalent attitudes towards science writers. Those interviewed often expect science journalists to be *au fait* with the essentials of their research areas, regardless of how abstruse or specialised these might be. Yet the same researchers occasionally object to critical questions from reporters; they feel that hacks never worked for years in their area, and are thus unqualified to criticise their work. (It is also commonly believed that journalists represent an extension of the public relations apparatus and exist primarily to promote science.) Some scientists, moreover, display ambivalence when they become peeved at science journalists who immediately grasp the essence of their research - as if journalists somehow trivialised aeons of hard toil by understanding their work so quickly.

Deadlines are another possible cause of friction between scientists and science writers. Journalists cover news on a comparative scale of Planck time, while many researchers work on what seems to be a geological, or even astronomical, time-scale. Coaxing some academics to meet media deadlines thus requires skill.

Deftness is also needed in handling the deluge of telephone calls, faxes, letters and emails from the general public and scientists inquiring about new scientific developments, requesting contact details for researchers and organisations and suggesting future news articles and features. The most persistent correspondents are invariably those proposing alternative explanations for phenomena that they feel defy orthodox science. Some editors consider it part of the science writers' lot to reply to most, if not all, such communications, be they from cranks or not. The matter does not rest there. Correspondents dissatisfied with science journalists' responses, or lack thereof, may directly approach editors who, while capable of evaluating many sporting or political objections on their own merit, are easily confounded by scientific protests - particularly those from senior scientists flaunting impressive letterheads and strings of letters after their names. With circulation wars rife nowadays, editors may further react to such letters of complaint as if they signalled the potential loss of subscribers.

Unofficially, the science reporter is regularly regarded as a repository of all knowledge related to science, technology, the environment and medicine (even when these areas are not covered). Out-of-the-blue questions thrown at me over the years by fellow journalists include: 'How much does the average frog weigh?' 'Are there more stars in the universe than grains of sand on "the" beach?' 'How many cells are there in the human brain?' 'How do you build a cell?' 'How does one store unused electricity?' 'Do atoms age?' and 'What do opals consist of?' This is in addition to a barrage of requests to perform - virtually instantaneously - a swirl of arithmetic or mathematical calculations, including those unrelated to science.

3. IDEAS

Although inevitably short of time and resources, not to say patience, the science correspondent is seldom at a loss for story ideas. When I open a book or read a news article or feature, ideas veritably leap out for follow-ups or other 'angles', the approaches taken by journalists to stories. The problem is not of originating ideas, as many people seem to think, but rather of finding the time, the appropriate scientists or enlightened editors to pursue them properly.

3.1. Sources

At least four sources of news and information are regularly used.

3.1.1. Agencies

A raft of news agencies, or 'wires' as they are dubbed, such as Reuters, Associated Press, Dow Jones, Agence-France Presse, United Press International and so forth, are available electronically on the desktop these days. In addition, several major British and American broadsheet newspapers - including *The Guardian, The Times, The Daily Telegraph, The New York Times* and the *Washington Post* - release syndicated stories and features to media subscribers.

3.1.2. Media Releases

Science writers seem to receive more mail - both physical and electronic - and telephone calls than other reporters. These communications stem chiefly from two sources: those wanting to promote their wares and, to a much lesser extent fortunately, those complaining about something that was written or not written (in the case of wares not promoted). The promotions come mainly in the form of media releases. Screeds of this American innovation, dating from 1907, are routinely emailed, snail mailed and faxed from universities, government agencies and research organisations around the world. Key international science journals, such as *Nature* and *Science*, provide embargoed releases every week, complete with researchers' contact details. Basing stories on such journal articles is a legitimate practice and not a form of plagiarism, as some editors seem to think.

Science writers dislike merely regurgitating press releases from other sources. Yet the insatiable demand for news is such that they sometimes resort to accepting releases as a primary source, supplemented perhaps with extraneous information and quotes from other scientists. Conducted wholesale and entirely uncritically, this practice may confuse the public understanding of science, a noble and worthwhile ideal, with the public support or promotion of science, which may result from organisations embarking on funding or recruitment drives.

3.1.3. Conventions and Contacts

Scientific conferences are another fountainhead of contacts and story leads. (On the few occasions when tightening newspaper budgets permit attendance at such events, however, science journalists are often expected to justify the expense by filing copious copy. This unfortunately leaves little time to listen to papers and forge contacts with delegates.) Personal contacts are also made by email, telephone and, all too rarely these days, over lengthy liquid lunches - which may encourage cagey scientists to loosen up.

3.1.4. The Web

I have found the Internet rather cumbersome to trawl every day for news stories. It's simply too time - consuming to visit the home pages of key universities and research organisations in search of new media releases or announcements that may make stories. Instead, the Internet is a useful reference source for stories already picked up by the wires or other media. For instance, if Reuters starts running a

NASA announcement about the discovery of a new planet, the space agency's home page might be consulted for further information, contact details, graphics or pictures. (So media-minded is NASA, incidentally, that it has been referred to as a media-relations bureau that runs a space agency.)

4. PRESS PRIORITIES

Before analysing what makes science newsworthy, it is worth mentioning that newspapers place their most prized stories on odd numbered pages, beginning of course with page one. The practice is based on what is believed to be a tendency by readers to pay most attention to 'right-hand' pages that face them directly; 'left-hand' pages are often read to one side while slowly being turned. In the layout of individual pages, however, the converse applies; lead articles generally (although not always) appropriate the left-hand side of pages because most people read from left to right. Readers also scan down rather than up pages. This is why the most cherished pieces tend to occupy the upper left-hand section of pages.

Sometimes, however, a story may be displayed prominently near the centre of a page in the form of a 'main block'. This is a bordered setup including a story, a photograph (frequently in colour these days) and perhaps a graphic or summary of the main points. The reasoning behind blocks is that surveys suggest readers glance first at the main picture on a page, and then read its caption. The orientation of photographs, meanwhile, indicates their perceived status by readers: vertical pictures imply that the subject is important 'hard' news, while horizontal ones suggest relatively 'soft' news.

The considered importance of individual stories is reflected in the size and style of typeface used. Key news stories tend to be printed in light type with large, 48 or 54 point headlines in bold. Lighter-hearted, off-the-wall pieces are printed in bold with smaller headlines in light type (often *sans serif*). The latter trend is more apparent in British Commonwealth nations than the United States.

That science stories not infrequently run 'light on bold' (light headlines with the text in bold type), with horizontal pictures, speaks volumes about how they are regarded by the media, especially in countries like Australia. Although the nation is no scientific backwater, the local media seem to regard science as something that happens abroad. Perhaps this is why science stories are often carried on foreign pages. 'If a story comes out of London or New York, they snap it up, even when it concerns Australian research,' notes the renowned British physicist and science author, Paul Davies, who now lives in Adelaide, Australia. 'This aspect of the cultural cringe is extremely dismaying for Australian scientists, many of whom are doing world-class research. I have known of occasions where I personally directed the Australian media to a local science story, only to have it ignored, and then carried a week later as an overseas news report.'

The problem is compounded by the fact that the Australian media seldom nurture specialist reporters. Not unlike politicians switching portfolios, journalists down under tend to swap beats with alarming ease, as though they would get bored with covering one round for more than a few years.

5. SELECTING SCIENCE NEWS

Editors invariably brim with confidence in their own objectivity. Their conviction is part of what may be described as a kind of 'media machismo', characterised by an authoritarian 'command and control' structure within most newspapers, a desire for order and certainty and a concomitant belief in absolute values. Yet the criteria for selecting stories are really quite subjective and vague and depend to a degree on the prevailing bias of individual media organisations. (Like some weather forecasters, editors are often in error but seldom in doubt.)

5.1. Criteria

Roughly ten criteria determine whether, where and how prominently science stories may be published.

5.1.1. Impact/Importance
This relates to several factors, including whether an event or discovery is unexpected. The headline 'Man bites dog', or the unlikely revelation that light travels at more than one speed in a vacuum, both meet this yardstick. An impact-laden story may also be unique or controversial or it could transform our understanding of the universe and our place in it, such as a possible future discovery by NASA scientists of a Martian meteorite that really did contain microbial evidence of extraterrestrial life.

Stories emphasising applications - especially lucrative ones - are generally preferred to those describing pure or fundamental research. (In order to get science stories into newspapers like *The Guardian* and *The Observer* in the early 1980s, before science sections came into vogue, I often wrote about commercial applications of science or technology for their business sections or 'city desks'. This trend towards commercialisation has returned with a vengeance today, giving technologists the whip hand over pure scientists.) Editors prefer applications with the potential to alter our lives - for instance the development of a cheap, ecofriendly fuel on which cars could run without requiring specially adapted engines.

Referring to such a fuel as a 'breakthrough' might further enhance its chances of attracting editorial attention. Many scientists wince at having their work, which may have built gradually on others' research, described in these glowing 'breakthrough' terms. I do not necessarily object to this practice. A touch of hyperbole assists in selling stories to news desks and often results in their being run more prominently and perhaps at greater length. Reservations about the full implications or nature of the research may be expressed lower down in pieces. This is the ideal place for such qualifications: senior editors seldom seem to pay much attention to material beyond paragraph three or four of science stories while many readers will persist to the bitter, reservation-ridden end.

One of the easiest ways to elicit a 'gee whiz' response from classically conditioned editors is to describe scientific discoveries or inventions in rose-red superlatives - the first, biggest, longest, tallest, fastest, cheapest and so on. This places a pronounced skew on science coverage. It's 'as though something was interesting only if it broke a previous record,' muses Brian Boyle, the Scottish director of the Anglo-Australian Observatory in Sydney. 'Imagine how little sports journalism there would be if that criterion were applied uniformly.'

5.1.2. *Relevance*

Impact relies partly on editors' fancies and whether they find stories important or relevant to their own lives. It also depends on the size and nature of readership. Editors question whether many readers would want to know about a particular story: is it entertaining or humorous, does it have 'human interest'? This may explain why the personalities and private lives of scientists, for example, are sometimes considered more relevant to readers than their research. Tabloid publications, in particular, find less appeal in stories dealing with complex, technical issues - say those of particle physics - than broadsheet papers. With newspapers competing fiercely these days with TV and Internet 'on-line news', however, numerous broadsheets are fast adopting tabloid tastes and content, if not smaller pages. Hence the popularity of 'sound-bite' quotes, the desire for which 'can frequently distort complex messages,' bemoans Brian Boyle, who nonetheless acknowledges their importance. 'Perhaps scientists need better training there too.'

The relevance of stories also depends on whether they are considered trendy or 'sexy'. This refers not to current subjects in scientific journals but to media whims. 'Topics like cosmology and genetic engineering are continually in the news because they are regarded as sexy, whereas condensed-matter physics or the chemistry of thin films hardly get a mention,' laments Paul Davies. 'In the end, however, the less trendy topics may be the more important. They are certainly less speculative.' He is absolutely right, but try telling that to a news desk.

Similarly, certain scientific figures - the British physicist Stephen Hawking and the Canadian-based environmentalist David Suzuki, to name but two – are deemed to be hot property. 'The curse of the celebrity means that a handful of media-savvy scientists are likely to be held in esteem far in excess of their true scientific status,' notes Davies, an international celebrity himself. 'In some cases, media personalities who are not even practising scientists are introduced as world experts. By contrast, geniuses such as theoretical physicist Ed Witten are almost unknown to the public. This distortion causes resentment among professional scientists and jealousy for the fame that prominent personalities may receive.'

5.1.3. Topicality

Put simply, topicality refers to whether stories are considered 'new'. News value is largely a function of time: the longer that stories continue, the less newsworthy they become. Thus the discovery that we live in an accelerating universe - which is expanding not at a fixed or slowing rate, as once thought, but at an increasing pace - may, at a pinch, make page one on its day of announcement but only page two or four the next day. After several days' coverage by other media, it is unlikely to be picked up at all unless a new angle can be found or concocted. Even then, it would probably be placed on page four or further back.

5.1.4. Exclusivity

Newspapers thrive on exclusives. Reporters announcing that they are onto a great story, will probably be beseeched by a drooling news editor: 'Do we have it on our own?' The credo 'first with the news' is uppermost in their mind, and journalists revel in forcing their competitors to 'play catch-up'. Some organisations, however, are reluctant to hand exclusives to favourite newspapers as they feel this alienates other media. Editors, too, may be cautious about accepting special offers. As one unusually wise and witty editor sometimes quipped when I produced exclusives, out of a hat, about unwieldy or technical topics: 'Do we have it exclusively because no-one else wants it?'

5.1.5. Background Information

This refers mainly to explainers and features whose prime purpose is to provide readers with sufficient detail to understand ongoing news events, such as landslides or earthquakes. Editors busy filling pages on such 'hot' topics may feel that explainers take precedence over new or important scientific happenings.

5.1.6. Utilitarian Value

Science stories are never more popular than when offered towards the weekend when news editors are searching sedulously for forward items to fill Monday's paper. Sundays are generally slow news days and so anything that is not hard news is often held over for what is known as the 'Sunday-for-Monday' slot. Some wily public relations officers specifically release news for these slots by slapping Sunday-midnight embargoes on certain stories. Similarly, science is welcome around Christmas and on public holidays (when scientists are generally unavailable) and other slow news days when little is happening.

Alternatively, the news 'mix' may call for a dash of spicy science to vary the usual dreary diet of politics, business, crime and sport.

5.1.7. Illustrative Potential

Science is well suited to explanation by way of profuse graphics and 'action' photographs depicting white-coated scientists brandishing their discoveries or inventions in laboratories, strewn with imposing equipment and instrumentation, or out in exotic places with dramatic backdrops. As they say, 'a picture tells a thousand words'. Graphics and pictures may influence decisions about where and how prominently science stories run. For example, news that biotechnologists had succeeded at last in engineering a blue rose would almost certainly rate a mention in most newspapers. Enticing colour photographs of the famous flower, and its creators, however, might secure the story as the main block on a page, if not on page one itself. Likewise, the use of dynamic, quite elaborate graphics may ensure that science hits the headlines.

Cartoons also frequently illustrate science stories, which are considered so incomprehensible by many editors as to be laughable.

5.1.8. Authenticity of Results

Editors sometimes wish to verify scientific 'breakthroughs' announced by companies. Findings released by universities and large research organisations, on the other hand, are accepted quite willingly. This is despite the burgeoning trend for such institutions to seek external financing from the same companies whose motives are questioned. It is hard - if not impossible - to dissuade news editors from publishing stories running on the wires, the broadcast media and probably the next day in opposition newspapers but that have not been published in peer-reviewed journals and simply sound dodgy. This is because editors, most of whom slavishly follow the pack, fear missing stories more than misinforming the public (whom they reason may be correctly informed later on).

For instance, the widely publicised claim that two chemists, Stanley Pons and Martin Fleischmann, had achieved nuclear fusion in a laboratory at room temperature was announced in March 1989 at an American press conference rather than in a refereed scientific journal. It turned out to be a red herring.

It is not just unpublished work that may prove unreliable. Even peer-reviewed research published in respectable international journals may later be questioned; such is the nature of the scientific process. In Australia, for example, Aboriginal artifacts found in a rock shelter at Jinmium in the Northern Territory were claimed by researchers to be up to 170,000 years old. The pronouncement made headline news in several leading newspapers and earned a Sydney reporter a prize for science journalism. A subsequent report dating the artifacts at no more than 10,000 years was largely ignored by some of the media that first reported the 'revolutionary find'. (The criteria used to bestow guerdons are worrying, as are some of the government agencies and organisations involved in funding them.)

Bert Roberts is an Australian-based earth scientist whose technique for dating individual grains of sand up to 800,000 years old overturned the original Jinmium findings. He argues that the peer-review process - itself under strain these days - is only the first step towards accepting scientific results or hypotheses. 'The most reliable science journalism is apt not to be the reporting of "cutting-edge" science–reviewed by only a few peers at most,' reckons Roberts, 'but rather those accounts of research published two or three years previously that have since been through the blowtorch of scrutiny by the wider community of interested but sceptical scientists.' By then, unfortunately, such accounts would have become, by journalism standards, science history.

5.1.9. Presentation

Ultimately, the way stories are presented to news desks may determine their fate. Editors possess notoriously short attention spans, so science journalists must carefully prepare their story lines beforehand. Long-winded, technical diatribes of scientific developments, particularly those delivered in a thin, halting voice, will likely receive a cold shoulder, as will recondite, theoretical treatises promising no gleaming pot of gold at the end of the rainbow. The winners are invariably easily conceptualised 'breakthrough' stories, announced in a stentorian, mildly euphoric but resonant voice, and loaded with superlatives, commercial applications, dollar signs and alluring picture and graphic 'opportunities'. This, after all, is the world of 'infotainment' (or, perhaps more aptly, 'entermation' because entertainment repeatedly takes precedence over information).

Science journalists forget this tenet at their peril, as I once did. A British astronomer, Jeremy Bailey, had found a clue in the ongoing search for the origin of life - a special kind of polarised light wave believed to give shape to the organic molecules that formed the basic building blocks of life. Although only part of the account of how life arose, the discovery contributed to resolving a mystery about the molecular structure of living plants and animals. It suggested that the manner in which life developed, although it may have originated on Earth, was determined in part by the way starlight scattered off tiny particles of dust found in giant gas clouds in deep space. Excited by the news, I dashed off to inform the news desk. But instead of simply stating 'scientists have found a vital clue to the origin of life', and leaving it at that, I catapulted into detailing some of the technicalities of the research. This is invariably a mistake, and the desk was duly unimpressed. It was suggested the story might make an article for my then weekly science page (considered a ghetto for technical trivia beyond the ken of average readers). I returned crestfallen to my desk. Later that day, I was approached by a frowning editor demanding to know whether a piece running on the wires about 'the origin of life' bore any resemblance to the one I had suggested earlier. 'It's the same story,' I replied, and was subsequently lambasted for not having explained it properly. 'We could have missed an important story,' he chided.

5.1.10. Knowledge Value

The longer that journalists cover science, the more they come to identify with scientists and the scientific process and the less they tend to relate to fellow journalists and the media industry. Hence most veteran science writers are excited by scientific developments that advance the frontiers of knowledge or provide some deep and genuine understanding of physical phenomena, irrespective of how esoteric they might be. Yet convincing news editors of the purpose of running such stories may be difficult, especially when the day's news list is groaning under the weight of an air accident, a cabinet reshuffle and perhaps a popular cricketer's resignation - especially when they all occurred locally.

Editors, like many journalists, are inclined to be sensation seekers who are addicted to the adrenaline surge brought on by a rousing story or a good outrage. For them, the ephemeral thrill of tracking down stories and riding high on the crest of waves of breaking news beats the more sedate, enduring satisfaction of science writers or academics seeking a deep understanding of fundamental phenomena or processes. It's unsurprising, then, that many editors revere thrilling, colourful or heart-rending tales, while versed science writers concern themselves with such mundane questions as knowledge value and validity. 'Emphasis on breaking news is often detrimental to good coverage of science, for important issues ... may not be associated with striking single events, and significance usually lies in long-term consequences,' writes Dorothy Nelkin, an academic and author at New York University, in *Selling Science*. Nevertheless, I think science journalists should alert editors to what they consider to be key developments, even if their counsel is neglected.

5.2. *Other Considerations*

Given the cornucopia of criteria for determining what constitutes news, what, in practice, does not? The answer, briefly: the subject matter of multitudinous media releases. That a university or research organisation is about to or may commence a new research project - irrespective of its size or expense - does not necessarily render the venture newsworthy. The world is full of projects in the making. Nor does the revelation that research on a major project is progressing, or that its results may be imminent, constitute news. Many such 'stories' are the products of organisations pushing for funding or justifying existing spending programs. Yet editors occasionally fall hook, line and sinker for such ploys, particularly when mainstream universities and research organisations sidestep science correspondents and approach news desks directly with local scoops. Once they get their teeth into stories, some editors are hard to deter - even when warned of their being 'old hat' or of dubious merit.

The foregoing criteria are employed not only by news editors selecting stories for the day's news list but also by those attending the daily news conferences. These are essentially two separate gatherings of the editor, news editor (and their deputies), some senior journalists and representatives from the pictorial and graphics departments. Morning conference peruses the day's news list and decides which and how stories will be pursued that day. The afternoon conference reviews an updated news list (based chiefly on reporters' progress), considers any latecomers and judges their respective merit - again largely using the above criteria. The editor then decides if, where and how prominently particular stories will be published. This is reflected on a page list indicating where individual stories will be placed.

Acceptance by news editors of a science story is clearly only the first in a series of precarious steps towards publication. It is crucial that news - desk personnel, who present stories at conferences, understand them thoroughly. Few comprehend much about science, however, and so it's not surprising that science stories are either dealt with cursorily, customarily in jocular vein, or simply read from the news list. All too often, the poor science story falls victim to 'Chinese whispers', and the night editors (who take charge after final conference) may expect something quite different - if anything at all - from what the science journalist proposed that morning.

That's not the end of it. Stories finally making it through the hoops are not necessarily safe. Last-minute hard news - a train crash, hijack or military coup, for instance - has the wherewithal to topple or displace other stories to lesser pages or positions. Considered fairly expendable beasts, science stories, needless to say, are sometimes the first to succumb. Even when they survive in one form or another for a newspaper's first edition, science tales stand an above-average chance of being dropped, or spiked, for subsequent editions.

6. CRAFTING THE PIECE

Writing science news should be distinguished from writing science features. News rushes quickly to the point, tells a story pithily (though not necessarily dryly), quotes a few authorities succinctly (quotes should state opinions rather than facts) and simply stops. Features, meanwhile, might explain the background to news events or describe affairs leading up to them. They luxuriate in a slow lead up, which whets the appetite and gently eases readers into what should be a pleasurable reading experience replete with colourful descriptions and examples, generous quotes from numerous sources and perhaps alternative explanations and perspectives. While there may be subtle variations, the basic format for news and features is broadly similar across newspapers and other periodicals such as magazines.

In a nutshell, the simplified and easily digestible science news story - comprising a lead, body and tail - is the antithesis of the highly detailed and impenetrable scientific journal article. Reverse the order of literature review, method, results and discussion, and you approximately have the structure of a news article. Journalists call this the 'inverted pyramid', in which paragraphs decline in importance from the top down. Science stories, however, lend themselves to telling in chronological order. So, while it's advisable to begin with the 'inverted pyramid', it is sometimes useful to drop this format after the first four or five paragraphs and to report the rest of the story in chronological order.

Opening lines are crucial. Tim Radford, the Science Editor of *The Guardian*, once reminded me: 'If you don't grab readers' attention at the start, there's no point in writing further.' (He also advised, sagely, that the key to science-writing success lay in carefully selecting subject matter and then writing about it well. Obvious though it may seem, this advice is not always followed.) The lead paragraph contains the key message and sets the stage and general tone. It should hook readers by firing the imagination and succinctly explaining the five Ws and an H: who, what, when, where, why and how. If this cannot be accomplished in 25 words or so, then the lead may be broken into two paragraphs or more. The lead should also emphasise applications, be they commercial or merely potential.

Personally, I prefer to launch the lead with a snappy sentence - the hook, or appetiser, as I call it - which somehow sums up the gist of the piece in a few witty words. A story revealing evidence of extraterrestrial life, for example, may begin with the hook: 'We are not alone.'

Needless to say, science scribes, like all writers, ought not to start writing until they have fully understood their subject matter. In practice, however, this does not always happen because some journalists are required to churn out two or more stories a day, and sometimes at the last moment.

7. WRITING STYLE

'Let the copy sing and dance,' Robin McKie, the Science Editor of *The Observer*, once instructed me. He was alluding to the use of rhythm, cadence and euphony, in addition to the ability to simply have fun with telling the tale. There are tips galore. They include the use of humour (where possible), short, simple words and brief but lively sentences and paragraphs. The golden writing rule is not to use several words when one suffices. For example, use 'consider' instead of 'in consideration of' or 'before' instead of 'prior to'.

Science writing lends itself to the liberal but judicious use of definitions and examples. Exercise caution, though, with analogies, some of which are as complex, or even more complicated sometimes, than the subjects they are intended to illuminate. In describing relative distances or dimensions, however, it's preferable to provide real-life analogies than actual measurements. For instance, the atomic nucleus might be described prosaically as being 100,000th of the diameter of an atom or, more vividly, as about the size of a fly at the centre of a football stadium. Dispensing with tedious figures, Paul Davies writes in *Ripples on a Cosmic Sea: The search for gravitational waves* that the hunt for these elusive waves requires 'detectors so sensitive they could pick up the equivalent of a pin being dropped on the other side of a planet. They must spot changes in length which, measure for measure, are the same as the width of a human hair in the distance to a nearby star. The vibrations they seek to monitor are so small, that even the noise of atoms moving about drowns them out.' That's science writing *par excellence*.

Perhaps the toughest question facing even the most seasoned of science journalists is what to omit. The answer depends in part on an article's length and the target audience, and hence the level of assumed knowledge and interest. As a rule, I assume that readers are intelligent but unfamiliar with the particular branch of science under discussion. I also try to refrain from using scientific jargon and excessive technical detail and avoid cramming in too much information, although the temptation is ever present. The aim is to strike a balance between overwhelming readers with complex detail, to which they cannot readily relate, and providing insufficient facts to enable a story to be sensible and accurate. A clearly written piece that inspires readers to consult the library or the laboratory for further information is better than one that is comprehensive but incomprehensible.

The overarching aim of science writing is to maximise simplicity and impact while minimising inaccuracy. This is achieved by reducing the technical detail of a story while emphasising its thrust and stretching potential applications to their credible limits. Simplicity may be checked with non-scientists and accuracy with scientists from a rival team.

An ostentatious or excessively flamboyant style should be avoided. In the English-speaking world, this tendency is particularly marked in countries like the United States and Australia where some writers seem to strive to impress readers with their skills or knowledge and their ability to produce lavish, deathless prose and a profusion of literary references and allusions. This is reflected in some headlines that may be rather ornate and smart at the expense of clearness and accuracy. The tendency also arises among journalists writing for the benefit of each other, or the judges of journalism awards, rather than for general readers. In my opinion, most science stories are complex enough in their own right and simplicity and clarity may be jeopardised by an extravagant style.

8. CHECKING AND EDITING

Publishing a science story without carefully checking the facts is like flying an aircraft without first checking the controls: neither readers nor passengers appreciate the effort saved. Facts may be checked and cross-checked in several ways, including in dictionaries, encyclopaedias, specialist reference works, books, journals, previous newspaper and magazine articles and on the Internet. In addition, scientists should verify facts and figures; better still, critics and opponents, who may be eager to detect errors in their competitors' work, should be consulted.

This raises the thorny issue of allowing scientists to peruse copy before publication. Many journalists frown on this practice which they feel undermines their independence. I disagree with this for two reasons. For a start, mainstream newspapers and magazines exist primarily to boost corporate profits by raising circulation, and hence advertising revenue. Thus the press itself, much of which is controlled by media barons and to some extent by corporate and political concerns, poses the biggest threat to independence. As Upton Sinclair, an American novelist and polemicist, is quoted as saying in *Objectivity and the News*, journalism involves 'presenting the news of the day in the interest of economic privilege'. Second, journalists need not comply with the demands of scientists who disregard the request to check factual accuracy only. This has seldom been problematic in my experience, as most scientists have kept to inspecting facts alone. Very rarely was I asked to alter a story's orientation or to actually expurgate information.

The editing process may also affect the accuracy of stories. After filing, copy is read by a series of editors who then forward it to subeditors, sometimes with recommendations or deletions. Subs, as subeditors are known in the trade, tend to be minutiae-minded folk who trim copy to the required length to fit on designated pages and, in the ideal world, rectify inconsistencies, mistakes in grammar and spelling and stylistic aberrations. They also strive to improve the readability of copy, and it is here that problems may arise. For example, the science journalist may have spent time carefully honing a description of some phenomenon or device that finally meets with scientists' approval. On reading the description, over-zealous subs may feel it can be improved - by which is generally meant shortened or simplified even further. Careful subs should check back such alterations with the reporters concerned, but sometimes they do not and the meaning of descriptions or concepts may become horribly distorted.

Further distortions may arise when haughty check subs, that monitor the work of subeditors, make amendments without consulting their subbing subordinates. In addition, senior editors occasionally alter copy already approved by subs and check subs. Finally, stories making it intact first time round risk getting distorted during production for a newspaper's second edition. Researchers would be less likely to blame science journalists for such mistakes if the text had been run past them beforehand. This is another reason to permit scientists to scrutinise copy.

9. MEDIA MANGLES

Scare stories abound. Elspeth Garman, a former nuclear physicist and now an X-ray crystallographer at Oxford University's Laboratory of Molecular Biophysics, recounts an episode with a British tabloid. The paper ran a story about the scientists placing reverse transcriptase, an enzyme involved in producing genetic material from the human immunodeficiency virus, HIV, aboard a space shuttle. The aim was to grow better crystals under conditions of microgravity. 'An out-of-control nuclear-powered Russian satellite was spiralling to Earth at the same time,' she recalls. 'The article implied that the satellite might hit the space shuttle and thus spread live AIDS virus and nuclear waste all round the globe. It was awful.' The paper had picked up the story from an accurate article on the microgravity experiments published in *The Daily Telegraph*. Putting two and two together, the tabloid had arrived at five without checking its copy with the scientists involved.

Headlines and occasionally picture captions - which are written by subs and not by science journalists, as many scientists assume - may be even more bothersome than efforts to 'improve' copy, as Garman once discovered. In January 2000, a prominent British newspaper interviewed her about the workings of a new influenza drug. 'Inventor of a vaccine to combat flu', proclaimed the headline across the resultant half-page feature. 'It isn't a vaccine and I did not invent it,' she retorts. 'I am so embarrassed I could crawl into a heap of leaves in my garden and not come out till next winter.' Fortunately, the text did not imply she was the inventor. 'But the headline is really bad for me, as it antagonises those who did invent it.' The journalist subsequently apologised - even though he probably wasn't responsible for the headline. 'He seemed to have no idea of the trouble so-called "inaccuracies" cause,' Garman laments.

10. SCIENTISTS SPEAK OUT

Are such incidents the exception rather than the rule? I asked several prominent scientists whether, in their experience, science writers had been fair and accurate in the way they quoted them or their research.

'Yes!' enthuses Brian Boyle. 'I have seldom come across a journalist who was unfair or inaccurate. Perhaps I have just been lucky. Where there have been inaccuracies, it has usually resulted from a breakdown in communication between the scientist and the science journalist.' He reminds that reporters 'frequently ask questions which cause you to view problems in a different light'. 'Science journalists can sometimes bring their own scientific training into play, which can result in a very interesting cross-fertilisation of ideas.'

'Almost always yes,' qualifies Bert Roberts. 'Only once to my knowledge have I been quoted in a way that was the complete opposite of what I had said.'

On the whole, Paul Davies, too, has been satisfied with the way science journalists handled stories involving him. 'However, in part, this is due to the fact that I write a lot of the material myself, and I always suggest that a journalist's copy is read back to me before going to press.'

Martin Rees, Britain's Astronomer Royal and a distinguished Cambridge University scientist, has enjoyed 'good and smooth relations with professional science correspondents in print media and radio'. But he has experienced 'less good relations - and often frustration or irritation - in dealing with more general reporters, and with TV, where there is often a strong pressure to sensationalise, and more scope for distortion during the editing process'.

Science journalists 'are a great asset', submits Robin Batterham, Australia's Chief Scientist, 'if they are knowledgeable of the target audience and possess the gift of translating incoherent enthusiasm into meaningful messages'.

Elspeth Garman now knows and trusts some science writers. 'Others I am not so happy about,' she adds. Bloodied but unbowed by a mangling media, she avers: 'Ethically, the public has a right to know how researchers spend research funds from the public purse. We thus have a duty to communicate our science in an understandable way.'

As an American sociologist, Robert King Merton, records in *Science, Technology and Society in Seventeenth-Century England,* 'science is public, not private, knowledge'. So say those remaining science writers who have not yet purchased a smallholding and relinquished the pen for the organic garden shovel.

11. REFERENCES

Davies, P. In D. Blair & G. McNamara, (Eds.) (1997) *Ripples on a Cosmic Sea: The search for gravitational waves.* Sydney: Allen & Unwin.

Blum, D. & Knudson, M. (1997). Field Guide for Science Writers. Oxford: Oxford University Press.
International Social Survey Programme. (1995-1996). National Identity Survey, *Australian Social Monitor*. Melbourne, Australia, 1996.
Merton, R.K. (1970). *Science, Technology and Society in Seventeenth-Century England*. New York: Fertig.
Nelkin, D. (1995). *Selling Science*, New York: W.H. Freeman & Company.
Sinclair, U. In D. Schiller, (Ed.) (1981) *Objectivity and the News*. Philadelphia: University of Pennsylvania Press.
Spinks, P. (1999). *Wizards of Oz*. Sydney: Allen & Unwin.

D. COHEN

11. PRESENTING A RADIO SCIENCE PROGRAM: ENGAGING THE PUBLIC INTEREST

1. INTRODUCTION

'You want me to talk about my latest paper on the radio? Without pictures, or diagrams, and no hand waving?' That's the kind of response radio science program makers receive when they call up researchers. And of course, everything the scientist says is true - there are no visuals on the radio. So what makes radio a good medium for communicating about science? How *can* you present science on the radio successfully?

Radio has one advantage over other media - it can respond rapidly to any breaking news story. Daily news programs regularly get to stories first - in a live program, all that's needed is to find an expert at the end of a telephone or, even better, persuade them into a studio.

With a well briefed presenter and a scientist who is a clear and concise communicator, even a short interview of a few minutes can inform the listeners of the essence of the development and its implications. A classic case of this was when the story broke of the cloning which led to Dolly the Sheep.

The most memorable radio has strong characters who tell gripping stories. Among the scientific community there are a number of individuals whose personalities come across well and who can relate their findings in an enthusiastic fashion.

Although it may sound surprising, pictures can get in the way in the explanation of scientific ideas. A TV item on a new discovery in cosmology demands images of dark matter, galaxies, stars or planets. The report of the appearance of genetically modified food will cry out for a picture of the food or the gene. But do stock pictures from the photo library of astronomical objects or shots of scientists walking around an observatory or sitting at a computer really tell the viewers anything about the research? In the second example, the classic visuals are of supermarket shelves and concerned shoppers, or of young female scientists poring over gels with dark and light bands or pipetting liquids into test tubes. Often the researcher can tell the story of the discovery more simply without the distraction of the pictures. A careful choice of words that describe a black hole or a galaxy can fire the audience's imagination better than many pictures.

Radio has another advantage over television in the amount of information it can put over. Curiously, the combination of pictures and words means that the density of information is lower on television than on radio. Words alone give a deeper insight with more of the subtleties of a discovery.

S.M. Stocklmayer et al. (eds.), Science Communication in Theory and Practice, 169–176.

2. THE NEWS STORY

So what are the different ways of featuring science and technology on the radio? The most straightforward approach is as a news story. Here the item is likely to be short - between 3 and 5 minutes - and presented as part of a general news program. In the UK, the interviewer will usually be a journalist/presenter without a science background. The story will have been selected because it has a news peg. An event will have just taken place - a conference, a press conference or a disaster with a scientific angle, say - or there will have been a paper published in a journal. Like newspapers and magazines, radio news programs respond to press releases sent out by journals such as *Nature* and *Science*. The way that a news program will cover such a story will be either to interview the author, or to create a debate between the author and someone else who, they hope, will hold an alternative opinion.

In news programs, scientists are often interviewed as if they are politicians - in other words, in a confrontational manner. The presenters' aims are to grill the researcher, rather than merely to elicit explanations. Some scientists are unhappy with this approach. An alternative approach by news programs might be to put the science story at the end of the running order, and to treat the research in a more light hearted way. This too can be disconcerting to scientists who are used to presenting their findings in a serious fashion to a group of their peers.

Science is by no means the only subject that is treated in these ways by news programs. In this way, it can be argued that the presenters are asking the scientists the questions the man or woman in the street would like to ask. The presenters do not pretend to be particularly knowledgeable.

3. FEATURE STORIES

In the UK, Australia and elsewhere stations such as the BBC and the ABC broadcast specialist programs that report on science and technology. Usually they are made by teams of dedicated producers and specialist presenters. They have a different role to play in the schedules from news reports and are directed at different groups within the audience.

The programs can take on a variety of styles. Some are magazines - they include a number of different stories, while others are single topic documentaries. Other forms are discussions or debates. A rarer style these days is the simple lecture or talk.

An example of a weekly magazine program made by BBC Science's Radio Unit is *Leading Edge*. In 1999 it was broadcast in runs of six. Each edition lasted for 28 minutes and contained 3 or 4 items. There was a main presenter - Geoff Watts, an experienced broadcaster who has been a science and medical journalist for many years - and a reporter, Tracey Logan. She, too, has spent her career as a broadcaster concentrating on science. There was one other regular voice, that of Roger Highfield, the science editor of *the Daily Telegraph*. His role was to put the findings in context and comment on their significance. He has a great deal of experience of translating difficult ideas into simple words in print - he is also very proficient at doing the same thing in speech.

So how is the content selected? Each item has a peg, in other words a reason for being included. A program that was broadcast in the middle of April 1999 featured a discovery of 3 planets around the star Upsilon Andromeda; a new idea about the origin of life on earth; a progress report on the *Foresight* program in the UK; and a study of an accomplished artist to investigate how his brain functions. The majority of items had a news peg - the planetary find had been published in *Nature* that week, as had the results of the investigation of the artist. At a physics conference, the government minister responsible for Science, Lord Sainsbury, had reviewed the work of the *Foresight* panels, which aim to identify areas where research money should be spent. The story about the research that suggested that 'the meteorites which landed on our planet four billion years ago could have seasoned the primordial soup with a virtual spice cupboard full of carbon-based materials' was slightly older, though this was the first time it had been covered on British radio.

Each of these stories had a different treatment. The observation of the new solar system was made by a researcher at San Francisco State University, Dr Geoff Marcy. The only way to get Dr Marcy on the air was to interview him over the telephone. As the quality of the line was not high, the interview ran for just a few minutes. The fact that the researcher was on the phone gave a sense of immediacy to the interview. The presenter's questions to Dr Marcy were straightforward and asked him what had he found, how had he done it and why was it so significant.

In the next story, the reporter took the lead. Her role was to guide the listeners through the story and her words were punctuated by short comments from several speakers. This technique is particularly effective when the program is explaining a complex idea or piece of work. The story in this case was rather complex - it was about experimental evidence for a hypothesis that a kind of chemical that is carried by meteorites could have started life on earth. The reporter's feature included soundbites from the chemist who had done the experiment and from one the pioneers of the study of the origin of life, Stanley Miller, who talked about the problem of understanding how life began on this planet. It also included comment from Harold Kroto, Nobel Prize Winner and expert in the chemistry of the particular group of chemicals, who stated that he thought the idea was plausible. The reporter's script provided the skeleton of the story, and the interviews gave the supporting evidence.

This style of feature can leave the contributors puzzled and even irritated. They are usually interviewed for some minutes - maybe 15 or more - but when they hear the final piece they discover that they appear for merely a minute or so. The other 14 minutes, however, have been paraphrased by the reporter or turned into questions that she will put to another contributor.

In this example, there was relatively little editing of the interviewee's comments. The other parts of the magazine were made up of edited interviews, punctuated by comments from the live guest in the studio, Roger Highfield. One of his roles was to make some general points about the close relationship between science and art, and how this piece of research was by no means the first instance of science being inspired by the visual arts. Thus, he provided a context for a piece of work. This can be crucial in making a scientific development relevant to a general audience that does not follow a partcular field nor read a weekly magazine like *New Scientist*.

For the third item, Geoff Watts had a few days earlier attended a conference in Manchester. He conducted interviews with Lord Sainsbury and with a chair of one of the *Foresight* groups, in the conference space. This is another way to bring a sense of immediacy to a program. His presence at an event also makes the listeners believe that he knows how the scientific community functions, and this gives him an authority. The audience can feel confident that Geoff Watts is well informed and knowledgeable about the subject matter.

Some listeners, however, are irritated by any kind of background noise, from the quiet conversation of a group of people to music. Producers are keen to use different sounds or recording space to provide variety and to make certain editorial points. But why use music at all in a speech program? A proportion of the audience dislikes it, and thinks it should be left to the many music stations. I would argue that there are some good reasons for playing music. The first is that it can set a mood - after all, television producers do this all the time. The second is to make an editorial point - the words of a song can be meaningful, or they can form a bridge between one comment and the next. Music can liven up a piece in which the speakers are dull.

The final item in *Leading Edge* had a lighter tone. As well as being about the workings of the brain it explored a human relationship - between the artist and the neuroscientist. So often, research is portrayed as the activity of a lone scientist or a group of people working in perfect harmony, as presented in the scientific paper. This piece showed the human side of research, though there did not appear to have been any friction in the collaboration. This story of how a scientist was studying a real life question is the kind of item radio producers love to cover - and we assume the audience, too, are interested in such practical reports. So much of science is not directly related to everyday experience. Some members of the audience are interested in science for its own sake - these listeners will listen whatever the story is. But the majority of the Radio 4 listeners is not particularly fascinated by scientific developments. We therefore need to find ways to sell stories to them. We believe we can draw them in by telling them why it is important for them to know what the researchers are up to. Some of the material will be purely of interest for its own sake. The most striking stories, however, will become the stuff of conversation around the tables, bars and offices, and will give the listeners more information and opinions they can accept or reject on subjects such as genetic manipulation of plants, or animals in experiments.

In the 1990s, as there were more and more networks fighting for audiences, broadcasting organisations have become acutely aware that they must know what their listeners and viewers need or desire. Research is sophisticated and, in Britain, it reveals how people use the radio and when they are available to listen to certain kinds of program. Recent analysis of the habits of the BBC's Radio 4 audience has shown that people with busy lives can rarely spare more than 30 minutes to concentrate on a single topic program. Another aspect of this multichannel world is that each network has very carefully to define its unique selling proposition, and then have a well thought strategy for branding itself. In the recent revamp of Radio 4, a Science Zone has been created at 9.00pm on weekday evenings. On the World Service, science programs too are 'stripped' across weekdays at the same times. These programs are broadcast four times each day. The audiences can now easily find the regular programs on both these networks.

The message from this increased awareness of the audience's behaviour is that anyone wishing to sell an idea to a network has to consider whether the subject matter is suitable for a particular slot, and the requirement of the audience at that time. For example, an analysis of the impact of quantum mechanics on the physics of the late twentieth century would be acceptable to an evening audience, which is perceived as taking a serious interest in science and having the time to sit down and concentrate for half an hour. It would not be acceptable to those who listen on a weekday morning. This group is searching for information based programs which do not tax them excessively and which they can dip in and out of if they are busy. A series about the crucial role played by climate change in the outcome of certain great historical events - for example, the impact of the cold winters of the 1940s on the outcomes of battles in the second world war - was commissioned by Radio 4 for broadcasting at 11.00am on four Tuesdays in the summer of 1999. Although these programs were made by a science producer, the narrative was driven by the history and the science took a secondary place. Nevertheless, I would argue that it is important for such programs to be produced by science specialists as they know which science has to be included and which does not, and they know how to translate difficult concepts into simple ideas.

The edition of *Leading Edge* that I examined in some detail above featured astronomy, neuroscience and art, the interface between chemistry and biology, and science policy. Other programs in this series dealt with particle physics, physiology, genetics, scientific archaeology, palaeontology and planetary exploration. Listeners often ask why some stories get much more airtime than others, and why is it that some areas of science and technology rarely receive coverage? People often point to chemistry and engineering as subjects that are poorly treated by the media in general. And radio is no exception.

For a story to be turned into a 5 minute radio piece it must have one strong central idea that can be easily grasped by the audience. Often basic research in all sciences seems unimportant to those outside the field - whether it is the working out of a pathway of a neurotransmitter, or the behaviour of a chemical species. To gain a place in a program research has to have relevance to the understanding of a disease, a new technology that impinges on everyday life, or a finding that makes people think about big questions, such as the origin of life, or the universe, the first humans, and so on. Physics and the biological sciences are more likely to produce research that allows such questions.

For anyone attempting to interest a producer in a piece of research it is essential for them to understand the editorial brief of each series. A story has to have a number of different facets if is to fill a thirty minute documentary. If you have a speaker who is taciturn he or she is not going to be booked to take part in a lively discussion.

4. DOCUMENTARIES

In recent years the Science Radio Unit has made documentaries on a variety of subjects with a variety of forms. *Frontiers* is a series which takes a piece of trailblazing research and tells the story of its discovery and puts it in its context. Some programs have the role of reporting on the science behind a health or environmental scare - the causes of antibiotic resistance, say, or the safety of genetically manipulated food. The program makers' aim is to explain the science–many other current affairs strands will look at other issues, such as politics or secrecy.

Other programs feature an individual whose ideas are challenging the orthodox and explore the research that supports this challenge. In this category Radio 4 has broadcast profiles of the Greek physicist who believes he has an accurate method of predicting earthquakes in the Aegean, and of a group of environmental scientists who have evidence from West Africa to support the view that the rainforest is not disappearing. Programs which air views that question the status quo can easily catch the attention of the audience, which is interested in novelty. I must point out that, although the researchers in these documentaries are questioning the received wisdom, they are not on the lunatic fringe. We believe that by giving airtime to them and to those who disagree with them, we can show the public the debates that exist within the scientific community. It is important that the listeners become aware that scientific knowledge is provisional and that ideas change as research continues.

The success of these documentaries turns on having a beginning that attracts the listeners, and keeps them glued to the radio. There was an ear grabbing start to a recent documentary about genetically modified foods. A more predictable opening might have been a series of comments from customers in a supermarketbut this has been seen and heard on the British media on numerous occasions. Once intriguing sounds have attracted the audience, however, it is not going to stay with the program unless the presenter quickly announces what the next half hour is going to be about.

The next essential factor is a wide-ranging interview with one or two central characters. These interviews become the backbone of the program, as they cover the entire story from start to end. The producer then selects the best comments and uses the remainder of the material as research for the linking script which moves the listener from one piece of interview to the next, from one part of the story to the next. To enrich the program there is other recorded material that fleshes out the story, or sets the context, or puts alternative views on the subject. Features like these are the most time consuming programs for a number of reasons. The first is that they can involve a great deal of original research, to find stories that have enough substance to sustain 30 minutes of broadcasting. Then it can take some days to track down speakers and fix up times and places to record the interviews. Second, these features often demand travel around the world to meet the researchers on location. Although this is sound only, an interview carried out in the field and face to face reveals a great deal more than one done over a telephone or ISDN line, even if the quality is very high. Direct communication is often essential if you are to gain the trust of an interviewee. A third reason is that the structuring and editing of the audio and the scripting of the linking comments can take up to a week. Ultimately it all has to fit into the allocated time slot, and often there is a final paring down of the program to achieve this target.

The weeks it takes to make a program like this may seem very short compared to the time allowed for a television feature. The time is determined by the budgets available for the respective media. Another difference is that nearly all the stages of the production of a radio program can be carried out by one person. A member of a BBC radio production team is today expected to research the editorial content, make the recordings themselves and then edit them, draft or write the linking script, and even sometimes voice it. They will hire technical staff - audio engineers - to mix the program. This makes the program making process very different from TV, though more and more of the technical side of TV production is moving into the hands of the production staff.

The formats I have described so far are complex; they involve many voices, location recordings and layered production. But there are two much simpler forms that can make gripping radio. One is the straightforward talk - one person telling a story or arguing a point. On the science theme, Radio 4 has broadcast 15 minute letters from scientists working in remote parts of the world. David Vaughan, a glaciologist with the British Antarctic Survey, wrote about the hardships of being cooped up in a confined space with someone who has a different taste in music and reading matter, 1000 miles from the nearest base. The aim of these letters is to show the human side of scientific endeavour; the personalities of the scientists and how they actually spend their time.

The other simple format is the conversation. Much of speech radio anywhere in the world is in this format, whether it's a DJ interviewing an artist about his or her new record, or a news presenter interrogating a politician. For several years in the middle 1990s BBC Radio 4 and the World Service broadcast a series of half hour conversations with leading scientists, doctors and engineers. The guests were chosen on a number of criteria - they had to have made a significant contribution to their field; they had to be engaging speakers; and they had to be prepared to talk about aspects of their lives as well as their science. Some scientists seem particularly shy about showing the public that they have interests other than their research.

A few of the guests were well known names world wide - Bill Gates and the palaeontologist Richard Leakey fall into this group. Some had carried out trailblazing research in medicine - Sir Richard Doll, whose work in epidemiology showed that smoking caused lung cancer, and Sir Roy Calne, the pioneer of kidney transplants. Sir Roy proved to be a fascinating interviewee because he is also a proficient artist who talked passionately about this side of his life as well as about his medical work.

From the evidence of responses from the Radio 4 listeners the most popular subjects were those whose lives brought science and religion together - this interest reflects the age and profile of the audience, which is on average in its fifties, relatively affluent and educated, and likely to be female. This is another case where it is helpful for the production team to be aware of the listeners' interests when planning a series. Again what makes the difference between a gripping interview and one that loses its way is a well-briefed interviewer. Carrying out research for this kind of program involves not just reading all the newspaper cuttings about the guest but also talking to their colleagues. In fact, often there are few cuttings about scientists as they are rarely profiled in the general press.

5. CONCLUSION

Any readers who have lived in parts of the world where radio does cover science on a regular basis will know there is one problem with this medium - it is very hard to find out in advance a lot of detail about programs. This is the disadvantage of being a rapid response broadcast medium. Much of the content of radio output is decided very near to the time of transmission. Although the basic idea for a program is developed to gain the commission in the first place, production teams have a matter of days or weeks at the most to put the flesh on the concept. By this time it is often too late to find written publicity for the programs. A further problem is that the newspapers in a country such as the UK give a small amount of space to radio in comparison with television, so even if a program is finished well in advance it is rare that it is commented upon or reviewed. Thus, while radio is a medium that devotes considerably more time to science, technology and medicine than television, and, as I have argued, is particularly well suited to exploring issues within these areas, many potential listeners are unaware of the range of programming made available by broadcasters such as the BBC.

This is beginning to change, in the UK at least, and in conclusion, I would urge future science communicators to remember that radio has a strong role to play in getting their message across to the interested public.

Radio Science, BBC, Broadcasting House, London UK

12. SCIENCE COMMUNICATION VIA TELEVISION AND THE WORLD WIDE WEB

1. INTRODUCTION

In 1995 I built the Australian Broadcasting Corporation's first website. It was for an information technology program entitled *Hot Chips*. As far as I can be sure *Hot Chips*[1] (notes at end of chapter) was the second TV program in the world to have a web site - the first being the BBC's *The Net,* also a program about computer technology.

At the time very few people in television had thought much about the Internet at all. It was perceived as an interesting phenomenon but maybe just another fad, like CB radio. Whatever the talk about its potential, the reality was that less than one percent of the population had Internet access and in TV terms, a one percent audience is just one notch above no audience at all. In those early days it was not at all easy to argue the case for why a TV show should have an on line presence.

As it happened, a few months before I began work on *Hot Chips* I had chanced upon George Gilder's 1994 book *Life After Television*[2]. In it he argued that the information revolution would inevitably lead to the end of television-as-we-knew-it. He also predicted that this would happen much sooner than anyone expected. At the time most of my TV colleagues found this idea simply incredible but Gilder's core arguments and reasoning have proved remarkably salient. By 1999, 21% of all Australian households had Internet access, the ABC was actively treating the Internet as an output medium in its own right, and the hot topic in TV boardrooms around the world was how to prepare for 'digital convergence'.

Digital convergence is a buzzphrase for the notion that you will soon be able to use one device to access any type of information or entertainment you wish. This notion implies profound changes to the conventional boundaries between differing mass media. What will it mean when consumers can use the same device to read a newspaper, listen to the radio or watch TV? What if that same device can also be used to make telephone calls, send mail and do shopping or banking? All of these things can already be done rather clumsily on the Internet. It is only a matter of time before Internet clumsiness gives way to seamless convergence. The coming communication revolution will affect every form of media.

1. TELEVISION

Most of us take television for granted. It is literally a part of the furniture of our everyday lives. The suggestion that television is under any imminent threat seems absurd only because when we think of television we think of what we use it for,

S.M. Stocklmayer et al. (eds.), Science Communication in Theory and Practice, 177–187.
© 2001 *Kluwer Academic Publishers. Printed in the Netherlands.*

which is mostly as a convenient source of news entertainment and 'infotainment'. We don't think of television in terms of what it actually is, which is essentially a distribution service offering only a very limited set of choices, rigidly delineated into pre-set timeslots.

Television has been the best available technology for the distribution of audiovisual news and entertainment for so long that it is easy to forget that it wasn't always so. Between 1930 and 1950 the cinema enjoyed a very similar status. In the then developed world, the treat of going to the movies was almost universally regarded as a regular part of weekly life. An evening at the cinema would almost invariably begin with a newsreel of the past week's events, followed by the main feature attraction for the night. Special matinée screenings were presented for children and one of the most common staples was a weekly serial. The advent of television didn't change this formula very much. It simply made it accessible via a more convenient medium.

This is why the digital revolution poses such a threat to television. The threat is nothing to do with people suddenly losing their appetite for news and entertainment and everything to do with a much more convenient and flexible way to obtain that news and entertainment.

In industry terminology this concept is known as 'video-on-demand'. It is about being able to watch whatever you want to watch, whenever you want to watch it, while at the same time enjoying VCR-type controls such as 'Pause' and 'Rewind' over the show you are watching.

The technology to do this has existed for a number of years. AOL-Time-Warner demonstrated a working experimental version as long ago as 1995. I was lucky enough to see it first hand and found it truly impressive. (I have subsequently seen a number of articles from various commentators describing this trial as a failure. Don't believe them. This trial proved both the feasibility of the technology and the concept). Video-on-demand is one of the most seductive and instantly appealing technologies I have ever seen. Cost and bandwidth are the only obstacles preventing its widespread introduction but not for much longer.

Since 1957 the cost-to-power ratio of computer hardware has dropped on average by 50% every 18 months. The phenomenon is widely referred to as Moore's law. While not a law in any scientific sense it has over the past 40 years proved to be an extremely accurate measure of the rate of progress in chip technology. Within the computer industry it is regarded as an almost axiomatic benchmark. It drives the R&D and investment strategies of all computer hardware and component manufacturers. The logic is self-fulfilling. If you know your competitors are likely to double their performance within 18 months and you want to stay in business, then the only course of action is to make sure your company is in a position to match them.

The other impediment is the provision of high-bandwidth to the home. The enabling technologies for this are well established and not particularly expensive. The major obstacle is not to do with technology or cost of technology but with politics. Telephone companies are in no hurry to deploy high bandwidth to the home because it implies a fundamental change to their business models[3] and is a direct threat to their ability to extract monopoly profits. Even so, the commercial pot of gold to be had from the provision of high bandwidth to the home is so great that even the monopoly Telcos will not be able to delay its advent for very long, at least in the democracies. If there is to be a bottleneck, it will be not be about technology but about politics.

1.1. The Achilles Heel of Television.

The inherent weakness of television has always been the indirect relationship between the viewer and the broadcaster. Because there has never been any way to make individual viewers pay for the parts of the service they actually watch, broadcast television is based upon a business model which forces viewers to pay indirectly. Most public broadcasters are paid for by taxation (or from a compulsory licence fee which is more or less the same thing), regardless of whether the taxpayer watches public television or not.

Commercial television is paid for by advertising, but the costs of this advertising are incorporated into the price of consumer products. Each time a consumer visits the supermarket or purchases any advertised product they contribute to the costs of commercial television, regardless of whether they watch it or not. Commercial television networks rarely acknowledge this reality. Their standard argument is that they do not charge the public so how can the service not be free.

However this indirect financial relationship between the individual consumer and the broadcaster is the Achilles heel of television. It places an upper limit on the revenue base. In the case of the public broadcasters, tax-payer funding will always be limited by the huge number of competing demands on the public purse. For commercial television, the unspoken reality is that there is a ceiling on the value of advertising to an advertiser.

2. SIZE MATTERS

The real product that commercial TV networks actually sell is the audience. In industry jargon, profitability is all about the number of 'eye-balls' or 'bums-on-seats' that can be delivered to an advertiser and with very few exceptions a big audience is much more profitable than a small one. This is the reason that commercial television is often described as a 'lowest-common denominator' medium. Programs for niche audiences simply do not help the bottom line.

The public broadcasters also face a pressure to deliver sizeable audiences. The size and demographics of their audiences are not the sole yardsticks that matter but they remain a very important factor when justifying the relevance and need for continued funding of a particular service.

In effect, both public and commercial free-to-air television are highly dependent upon the ability of the medium to attract a mass audience. If for some reason this mass audience were to sharply decline, it would mean a total collapse in the financial viability of the existing free-to-air business model.

Since the early 1980s the audience 'share' for free-to-air television has been in a slow but steady decline. This decline has been driven largely by the advent of alternative forms of entertainment such as the VCR, the home computer and pay and cable TV. This small decline has already impacted on the bottom lines of commercial TV networks because smaller audiences directly translate into lower revenue from advertisers. Most TV networks have responded with a very hard-headed approach to containing or reducing the costs of local content and by placing a much greater emphasis on amortising the costs of quality programs via co-production arrangements with international partners. For independent program-makers this has translated into a much tougher environment in which to make a living.

Yet the decline in the free-to-air television audience thus far is trivial compared to what is likely to happen over the next ten years.

Half a century ago the introduction of television ended the golden years of cinema. In one stroke television took away the mass cinema audience, not only devastating the revenue stream from the box office but at the same time dramatically increasing the cost of producing films that would attract people back to the cinema (wide-screen epics, colour, even social-marketing efforts such as drive-in theatres). It took decades for the cinema industry to recover and it will never again enjoy the dominance it had prior to television.

Something similar will happen to television over the years before 2010. Free-to-air television will survive, but only by learning to concentrate on the things it does better than any other distribution medium, just as radio and the cinema have done. The golden years of television should be thought of as what they were - a product of the best available technology at a particular time in history.

2.1. Life After Television

Ultimately, the new era will be highly beneficial for program-makers.

Under the existing system the free-to-air television networks act as gatekeepers between the audience and program-maker. In order to get your program on TV you currently have to sell it to one of the networks. In fact, if you can't sell your program to one of the networks it is very unlikely that you will be able to raise the finance to produce it in the first place. In negotiating this all-important sale, program makers inevitably discover that the networks hold all of the aces because they control the access to the distribution medium.

In the coming era of video-on-demand this monopoly will be broken. The financial viability of a non-drama project will no longer be determined solely by the 'deal' that can be arranged with a TV network. Instead, free-to-air TV will be just one of many distribution outlets. This will represent a major shift in the economics of non-drama filmmaking and will be a shift very much to the advantage of the program maker. It should enable a much more diverse range of programs to become economically viable and this will be a very good thing when it happens, however the immediate challenge facing program-makers is how to survive between now and then.

3. GETTING THERE FROM HERE

It is already quite tough to make a living as an independent filmmaker. The amount of money required to produce a quality program is substantial, and the days when broadcasters were prepared to fully fund non-news programs are receding. The trend is for networks either to purchase only the national rights to a program, leaving the filmmakers with the problem of recovering their budget by selling the program to the rest of the world, or else to seek to make the program as an international co-production with other broadcasters. This is true for most forms of television content apart from news.

At first glance it would seem that science and natural history programs should cross cultural barriers more easily than, say, drama productions and for that reason they should more readily lend themselves to international co-production. In practice international science documentaries appear to be limited to the staples of animals, sex, space, weather, disasters, dinosaurs and 'freaky' people.

Establishing a co-production between different broadcasters is never an easy task. Each potential partner comes to the table with different scheduling and audience expectations to attend to. Is a one hour program 50 minutes or 60 minutes in duration? How do you accommodate a partner who only has a 30 minute slot? Who controls the editorial? Is the program primarily infotainment or is it discussing issues of topical relevance, for example, genetically modified food? International audiences can have quite different expectations of what to expect from a science program.

Then there are the toughest questions of all–how should the costs be shared and what ancillary rights should each of the participants obtain in return? Who will provide the production team and facilities?

This last question is very important for aspiring science-communicators. It involves a type of Catch-22. No matter how gifted or talented you are, international co-producers will not support a project without being convinced that the creative people involved have a proven track record. Yet how does one develop an international track record?

Unfortunately there is no standard answer to this question. If you ask people who are working in the field today, you will find that almost all of then arrived by a different path. The one thing they have in common is that very few of them were working in the field 15 years ago. It is very likely that in 15 years there will be a completely different set of people making science programs. How will they get there? The media revolution we are witnessing is full of uncertainties. There is no road-map.

4. TELEVISION AND NEWSPAPERS AND THE WORLD WIDE WEB

The differences between these three media forms are profound.

i.   Television and newspapers are inherently one-way forms of communication. The Internet is a both-ways and many-ways form of communication.
ii.   Television and newspapers are very expensive businesses with very high barriers to entry. Compared to other forms of media the Internet is relatively inexpensive and there are few barriers to entry.
iii.   Television and newspapers can reach very large audiences simultaneously. The Internet is not particularly good at reaching mass audiences simultaneously (although it is improving all the time).
iv.   Television content is ephemeral. Newspaper and online content can remain available 'on-demand'.
v.   Television networks and newspapers are organised on a local/national level. Audiences are limited to the choices within their TV reception range or by the reach of a newspaper's distribution network. By contrast the Internet is a global medium. Audiences can obtain content from anywhere in the world.
vi.   Television networks and newspapers rarely refer to their competitors. Internet competitors often find advantage in linking to each other.
vii.   Television is overwhelmed by 'lowest common denominator' culture. Newspapers and Internet are much more a 'first choice' culture. The audiences actively select content and are never in 'blob' mode.
viii.   Audience 'flow-control' strategies similar to those employed by television networks have had only limited success in the online medium. Marketers who talk about control of Internet 'eyeballs' have failed to understand that the Internet audience is not in TV 'mode'. The online audience has the power to determine its own agenda. It is not a collection of easily shepherded 'eyeballs'. This is part of the reason why this new medium is so exciting.

5. THE INTERNET PHENOMENON

The Internet draws its power from three distinctive characteristics.
    The first is that it is a truly global medium. Wherever you are in the world, once you have access to an Internet connection you have access all billion pages[4] of the World Wide Web. Web-sites located on the opposite side of the world are only trivially more difficult to reach than those in your local neighbourhood. There is something quite wonderful about being able to do this - a thrill which I'm sure is not unlike that felt by the people who built their own crystal set radios in the 1920s and found they could hear a broadcast from a town 80 kilometres away. The Internet means the tyranny of distance need no longer be an impediment to information flow.

The second factor is that the Internet is completely digital and enjoys all of the characteristics of the information economy. Unlike traditional manufacturing, the expensive part of information content is almost entirely in its original creation. Replication and distribution costs are very near zero. Yet while original content is always potentially valuable, the realisation of that potential depends upon the content finding an audience. For this reason it makes sense to share or swap or even sometimes give away digital content. The goal becomes one of attracting attention and establishing credibility within a global smorgasboard of choice. Content is very important but is only one part of the success equation.

The third distinctive characteristic of the Internet is the most obvious one. It is a network of networks - of not only computer networks but also of human networks. It is a communication tool. Unlike traditional mass-media, the Internet is tailor-made for what in television terms used to be called narrow-casting - in other words specific content for specific interest groups. This is a challenging phenomenon for the traditional media because specialist interest groups can now create and operate communication channels which are completely independent of TV, radio and newspapers.

The more excitable evangelists of the online medium often speak in broad-brush terms about how the Internet is going to change the entire world media order. While it is true that the technology is going to change the nature of free-to-air television, it is quite wrong to think that the existing media organisations are going to somehow disappear or become irrelevant. The most likely scenario is that they will exploit their assets of expertise and brand-name credibility and adapt (admittedly with some pain) to the new media environment.

At the moment there are very practical reasons why the conventional mass-media are not particularly good at covering specialist areas such as science. A TV or radio news bulletin is a small capsule of information that needs to be all things to all people. This means the coverage is inherently generalist - a bit of news, a bit of politics, a bit of sport, a bit of weather. For a science story to be included in a bulletin, it needs to be of universal interest or else playable as a 'colour' piece. (Newspapers suffer from a variation of this same problem–although in their case it is not time but space which is the limiting factor).

Most media newsrooms are aware of many more interesting stories than they could possibly publish or broadcast. Some stories are killed on the basis of inadequate quality, but a large number are filtered out simply by the demands of the particular output medium. This is where the online medium becomes very interesting. Time and space need no longer be restrictions. A bulletin no longer needs to be all things to all people. It becomes possible to cater to specialist audiences. For example, an online news service can safely create a section purely about science without fear of alienating the part of the audience not interested in science.

ABC Online News operates a *Sci-Tech*[5] section in exactly this way. All of its stories come out of the central ABC newsroom and are generated by the normal news gathering process. Normally only a few of these stories would make it into a

TV or radio bulletin and without the online medium they would have been discarded for want of an outlet. This is one example of how media organisations are beginning to adapt existing resources to generate online content at very little cost.

The ABC also has *The Lab*[6] which is the online gateway to all ABC science coverage. This is a specific Internet initiative which functions as an output stream in its own right, generating original online content which not necessarily derived from any of the ABC's broadcast activities. Among other things, *The Lab* has its own *News in Science* service. *News in Science* stories tend to be longer and more detailed than those in the *Sci-Tech* service–more like a meal than a snack. (This difference seems to be a by-product of the medium for which a story is originally written). The point about *The Lab* is that it is an example of a mainstream media organisation adapting its existing communication expertise to a completely new medium. The people who created it applied communication skills honed from a great deal of experience in television and radio.

6. SUCCESSFUL ONLINE SCIENCE

The scientists who invented the World Wide Web[7] were in pursuit of ways to improve their communication with other scientists. In spite of the reality that the modern Web is being employed for uses far removed from this original purpose the underlying architecture has remained essentially the same. The medium is intrinsically suited to the cooperative exchange of information and ideas.

Popular wisdom about the behaviour of the online audience says that most people ultimately settle into a pattern where they tend to regularly visit a maximum of about seven Internet sites. This is the reason why there has been so much emphasis on the idea of building so-called 'portal' sites - sites which aim to be gateways to everything you might want to do online. Each of these portals attempts to direct its 'captive' audience to related sites within its 'family' and thus maximise the portal's total traffic and advertising revenue. The entire idea is derived from the television model of control of 'eyeballs' and depends upon the audience being ignorant of their power to choose where they want to go.

7. GLOBAL COMMUNITY

This 'portal' interpretation of popular wisdom doesn't necessarily relate to reality. If our experience with *The Lab* is any guide, science communication sites around the world are only too happy to link to other quality science communication resources whenever they become aware of them. The world's science websites are not in any formal network as such but they share a common interest, and in many ways the science culture of connecting ideas. If you have built a website which adds to the world's pool of science communication resources, you should make a point of alerting other science sites. There's a very high likelihood they will include you in their link list.

There are also excellent email resources available online for science communicators. If you are building online science resources you should at the very least make a point of ensuring that the science communicators within your country are aware of what you are doing. They will give you valuable feedback and may help you to spread the word. These email lists can also be an excellent sounding board when you want to 'privately' test a new web idea.

8. INNOVATION

One way to attract attention to a website is to be innovative. The online medium is still blue-sky territory. It is full of potential for innovation. If you can come up with an idea that has never been done before you can attract considerable attention. There are a host of net-guides looking for material to talk about. Being first with a 'cool' idea can generate a lot of interest. In 1998 *The Lab* hosted a year-long online diary from one of Australia's wintering Antarctic expeditioners[8]. This wasn't the first online diary from Antarctica, but it covered the whole year as it unfolded and included fabulous pictures. It attracted a world-wide audience. Never underestimate the power of a well-executed good idea.

9. DIRECT NETWORKING

The importance of email alert lists should also never be underestimated. Invite visitors to your site to subscribe to an email alert about future updates. Be sure to state a clear privacy policy and rigidly stick to it. No smart person will subscribe to your list if you do not.

It is relatively easy to set up an alert list - these send one-way email messages to people who have expressed an interest in hearing about updates to your web service. They are an excellent way to alert people whenever you publish something new. It is much more difficult to establish a successful interactive 2-way discussion list but if you can do so you will have enabled a cyber-community of friends to the basic theme of your site. There is no point in trying to establish an interactive-email list if one already exists which covers the same theme. It is much better to join an existing list and let them know what you are doing online. This is the power of the Internet - it is not in ownership but in networking.

On the subject of email, you should endeavour to reply to everyone who writes to you via your website address. Treat the fact that they have written to you at all as a compliment, even if the messages are sometimes critical. You can learn a lot from email.

10. CROSS MEDIA

You may be publishing online but that is no reason not to use other media to get the word out. Newspapers, net-guides, computer magazines and even TV and radio can be very useful means of promotion. You probably do not have the budget for advertising, but there is always the chance of having your site come up as part of the normal editorial content of one of the conventional media. Send press releases whenever you have something newsworthy or interesting. There is a huge amount of interest in the online medium so if you have a good story to tell, the conventional media may well want to talk about it. Often times you will be able to get on the radio talking about a hot topic and the only credit you will get will be 'this is xxx who is from the yyy website', a pretty good bonus because you have also just been on the radio talking about science.

11. SEARCH ENGINES ARE YOUR FRIENDS!

Every time you publish new material, you should not only alert the Internet search engines but ensure that the web pages include the Meta-tags[9] to help the search engines index your content. Search engines can take up to a month to display your content so they should be thought of as only being of benefit in the longer term, but they can be a valuable way to introduce completely new visitors to your site, especially if you've got specialist content.

12. A FINAL WORD

People often ask me about what sort of skills are needed in order to begin an online career. The best advice I can give is a variation of the old adage for prospective writers. 'If you want to be a writer, be a reader'. In other words, if you want to create great web sites, you should spend a good deal of time exploring what other people are doing online. You will develop a sense of what you like and do not like, about what works and what does not, about what is exciting and what is not. When you start creating your own online content you will find that all of this 'reading' has helped develop your own sense of style.

But be prepared to learn a completely new web-tool or skill or concept every 3 or 4 months. This is the rate at which new online technologies are unfolding. The upside is that the ultimate potential of this new medium is still very much unknown. We are in uncharted territory Very few generations are ever presented with the opportunity to participate in the early stages of a technological revolution as big as this one.

iana@your.abc.net.au

13. NOTES

1. Archived version online at http://www.abc.net.au/hotchips

2. Life After Television. W.W. Norton & Company ISBN 0-393-31158-9

3.'If you take the example of video on demand where you see the movie you want to see when you want to see it, the price might be a few dollars above the video rental store. Now for that we need to show you a movie of one and a half hours - which is roughly thirty times as long as a telephone call, we need to use between one and two million bits of information per second which is typically say six hundred times more than a telephone call. Put it together and it's eighteen thousand times more information and we're talking about a price that's maybe only four or five times higher than a local telephone call. So what you see happening is an absolute paradigm shift in the delivery of information. This is a fundamental change and it will bring with it fundamental changes in the way in which communications is packaged and sold'. Bob   James.   Telstra   Strategic   Development   (Quote   recorded   in   1995).   See http://www.abc.net.au/http/pipe/nextbig2.htm

4. 'It's a small Web after all' - News In Science, Sept 10, 1999

http://www.abc.net.au/science/news/stories/s52316.htm

5. Sci-Tech   http://www.abc.net.au/news/science

6. The Lab   http://www.abc.net.au/science

7. Strictly speaking the World Wide Web is a sub-set of the Internet. In practice if you work with one you inevitably work with the other. People often use the terms interchangeably. It is not worth losing sleep over the distinction

8. Ingrid On Ice   http://www.abc.net.au/science/antarctica/ingrid

9. There is a wealth of online information about Meta-tags. For up-to-date information you should consult your favourite search engine. At the time of writing, (2000) www.google.com was a very good place to start such a search.

SECTION 4

SCIENCE COMMUNICATION IN PRACTICE:
CASE HISTORIES

P. BRIGGS

13. NEW VISIONS FOR ASSOCIATIONS FOR THE ADVANCEMENT OF SCIENCE: A CASE STUDY

1. INTRODUCTION

In 1991, organisations based in 28 countries around the world agreed to establish an international federation of associations involved in promoting public awareness of science and technology. Many of them owed their existence in some way to the British Association for the Advancement of Science (the BA) founded 160 years earlier. In order to survive through that period, the BA and its sister bodies have had to face up to significant change both in the world in general and in science and its institutions in particular.

Challenges and opportunities continue to confront all associations, which must grapple with them if they are to remain relevant and viable into the 21st century. This chapter focuses mainly on the BA and how it is responding to its own situation but reference is made to other associations wherever appropriate.

2. ORIGINS

In 1830, Charles Babbage, best known as the father of the computer, published his *Reflections on the Decline of Science in England.* He was not alone in his views of the state of British science. At about the same time David Brewster wrote in *the Edinburgh Journal of Science*, of which he was editor:

> Bribed by foreign gold, or flattered by foreign courtesy, her artisans have quitted her service – her machinery has been transported to distant markets – the inventions of her philosophers, slighted at home, have been eagerly introduced abroad – her scientific institutions have been discouraged and even abolished – the articles which she supplied to other states have been gradually manufactured by themselves...

Brewster's sense of lost opportunity and of complacency on the part of the scientific establishment stimulated him to further action. He persuaded the Yorkshire Philosophical Society to host a meeting in September 1831 to consider setting up a 'British Association for the Advancement of Science'. The 250 people who attended agreed to do so and adopted three objectives for the new Association:

i. to give a stronger impulse and a more systematic direction to scientific enquiry;
ii. to promote the intercourse of those who cultivate science in different parts of the British Empire with one another and with foreign philosophers; and
iii. to obtain more general attention for the objects of science and the removal of any disadvantages of a public kind which impede its progress.

S.M. Stocklmayer et al. (eds.), Science Communication in Theory and Practice, 191–201.

In the ensuing years, the BA fulfilled these purposes through a variety of activities. It discharged its international obligations in two ways. One was to hold some of its annual meetings in other parts of the British Empire–Canada (in 1884, 1897, 1904 and 1924), South Africa (in 1905 and 1929) and Australia (in 1914). In addition the BA stimulated the foundation of similar associations in many countries. The American Association for the Advancement of Science (AAAS, 1848), the Australia and New Zealand Association for the Advancement of Science (ANZAAS, 1888), the Indian Science Congress Association (ISCA, 1888), the South African Association for the Advancement of Science (1903), the Canadian Association for the Advancement of Science and others were all inspired by and/or modelled on the BA.

3. EARLY YEARS

The BA was founded at a meeting and the Annual Meeting became its principal activity and a characteristic of all the bodies to which it gave rise. The meeting was held in a different city each year, something said to be inspired by the circuit of towns and cities undertaken by the Merchants of Light of Salamon's House in Francis Bacon's *New Atlantis*.

In the 19th century the BA Annual Meeting was an occasion for the birth of new terms, for the announcement of discoveries and inventions, and for major debates. For example, the use of the terms 'scientist' and 'dinosaur' was first proposed at BA meetings by Whewell and Owen in 1834 and 1841 respectively. Joule described his experiments on the mechanical equivalent of heat in the 1840s, Bessemer his steel process in 1856, and at the 1894 meeting Ramsay announced the discovery of argon and Lodge gave the first demonstration of wireless transmission. At the 1860 meeting in Oxford, a discussion after a paper in the Zoology Section turned into a confrontation between Samuel Wilberforce, the Bishop of Oxford, and T.H. Huxley over Charles Darwin's ideas on the origin of species. The clash became legendary - the best known 'victory' [for Huxley] of the nineteenth century, save Waterloo. according to Darwin's biographers Desmond and Moore.

Through the work of various committees, the BA concerned itself with or became responsible for many aspects of scientific and technical affairs in Great Britain. For example it played a major role in establishing electrical standards and even defined standards for the threads of screws!

4. TODAY'S CONTEXT

4.1. Mission Today

Gradually, the rise of specialised scientific societies, a process to which the BA often contributed, and the advent of scientific journals meant that scientists no longer waited to announce their results at a once-a-year meeting. Their first priority was to attend the meeting of the society most relevant to their own subject. No longer was the BA the only medium through which scientists could debate their concerns. The Association's role as the 'Parliament of Science' no longer had the importance it once had. In order to survive the Association had to change.

Over the years the BA has gradually focussed its activities. No doubt Charles Dickens, who satirised it in 1838 as 'the Mudfog Association for the Advancement of Everything' would be delighted! Today its mission is

> to promote understanding and development of science, engineering and technology throughout the UK, and to illuminate and enhance their contributions to cultural, economic and social life.

4.2. Science communication in the UK

Three documents have provided crucial stimulus to the development of science communication in the UK since 1985. In that year the Royal Society (the UK's Academy of Science) published a report on *'The Public Understanding of Science'*. Widely known as the Bodmer Report, after the chairman of the committee that produced it, the report became a landmark document. Above all, it led to a significant change in the attitude of the scientific establishment towards the public communication of science. Its importance in legitimising–indeed, *encouraging*–the involvement of scientists in communicating with the public cannot be overstated. The report led to a surge of activity, including the establishment of COPUS, the Committee on the Public Understanding of Science, which was set up jointly by the Royal Society, the Royal Institution and the BA.

The Bodmer Committee was established as a result of a recommendation of an earlier Royal Society committee on science education. At the time there were many science curriculum development initiatives in the UK and it was recognised that, to be effective, curriculum reform needed to be matched by improved awareness and appreciation of science in the wider community. Ironically, the Bodmer Report might not have been needed had the BA been doing its job effectively in the 1980s, an interesting reversal of the situation in 1831 when the complacency of the Royal Society prompted the formation of the BA.

The second key document was *Realising Our Potential*, a government science policy statement published in 1993. This provided official recognition from government of the importance of public understanding of science and promised resources to support activities designed to improve it. Importantly, it ensured that the UK's research councils, which fund scientific research, were obliged to promote public understanding of the sciences in their remit.

Most recently the Science and Technology Committee of the House of Lords, the second chamber of the UK parliament, published a report on *'Science and Society'* in 2000. Against a background of an ever-increasing number of science-related issues of major public concern, the report recognised that public trust in science and scientists would be a key issue at the beginning of the 21st century. One of its major recommendations was that communication between scientists and the public should increasingly become a two-way process. The report stressed the need for a 'culture of dialogue' and for scientists to use their ears as well as their voices in the course of engagement with the public.

The report, which largely reflected the evidence given to the committee, continues to provide an opportunity to take the science communication 'agenda' in the UK forward. 'Public understanding of science', so hotly pursued and written about since the mid-1980s, is rejected as a dated and somewhat patronising phrase to be replaced by the language of transparency, openness and dialogue.

4.3. The Growth of a Movement

There is a long tradition of communicating science to the public in the UK. Until the mid-1980s it was essentially the province of a small number of national organisations, local bodies and enthusiastic individuals. In addition to the BA, those involved nationally were the Royal Institution, founded in 1799 and dedicated to both scientific research and public communication of science, and the Science and Natural History Museums, both established in the South Kensington area of London following the Great Exhibition of 1851.

Today, however, the number of organisations that play an active role in science communication is vast. They include scientific societies, led by the Royal Society, engineering institutions, research councils, universities, science museums and science centres. These centres, which were unknown in the UK until the mid-1980s, have grown in number and size. Thanks particularly to the UK's National Lottery, a number of large-scale science centres opened in different parts of the country in 2000/2001.

Yet although many bodies are concerned with science communication or public understanding of science, there is no common understanding of what these terms mean. The phrases are in effect *banners* under which individuals and organisations with many different purposes can conveniently come together. They may exist to lobby for more funds for science; to encourage young people to pursue careers in science; to show the role of science as part of culture; to empower people to take part in discussions on science-related public issues; to ensure that people can make informed personal choices; or simply to provide science-based visitor attractions. The increasing number of individuals involved, the range of organisations for which they work and the variety of purposes for which such bodies exist can perhaps best be described as constituting a science communication *movement*.

5. THE BA TODAY

5.1. Role and Activities

The BA consequently operates today at a time when there is widespread interest in its purposes but also considerable competition for audiences and funds. Its reaction has to been to create a distinctive set of activities, to position itself as a key player, and to be seen as a hub for some of the national endeavour. Its work focuses around the UK's National Science Week, the BA Festival of Science activities for young people and science communication projects.

5.2. The National Science Week

Nowhere are the roles that the BA sees itself playing displayed more clearly than in the National Science Week. The first such week was held in March 1994 and immediately hailed as a success. By 2001 some 1500 organisations, institutions and groups were arranging thousands of opportunities for people to get involved in activities in cities, towns and villages throughout the country. Over 1.5 million people are estimated to have done so. The BA co-ordinates the week, puts together a national program, and liaises with the media. It supports organisers of activities by arranging workshops around the country and producing guidelines on event organisation and publicity. Schools, museums, companies, universities, local societies, hospitals, shopping centres and others all get involved. There is extensive national and local media involvement, with features and special programs as well as coverage of activities. Funding for the BA's co-ordinating role is provided by the Government's Office of Science and Technology (OST), which also funds a small grants scheme to support activities during the week.

The main factors in the success of the week have been the open approach taken by the BA to the program, with little or no attempt to exercise control over the contents; the large number of local and national organisations and enthusiastic individuals eager to run activities; the interest and involvement of local and national media; and the moral and financial support provided by government.

As a result the week is bigger than the sum of its parts and benefits all its stakeholders - activity organisers, the BA, Government and the public.

The nature of the week means that its audiences and purposes are broad, spanning those of the individual activities. Many, but by no means all, of them are aimed at young people. A significant feature of the week is the growing number of events that take science to the public through displays and activities in shopping centres and other places with general public access. In this way, the week impacts upon many who are not reached by conventional science communication activities, the majority of which involve attentive audiences.

Science weeks and festivals exist throughout the world. The possibility of establishing a UK science week had been discussed for some time before the BA took the initiative in 1993. By doing so the Association established its claim to be a key player in this area in the UK today. In other countries the picture is different. Where associations do not take opportunities to change and develop, where they do not take advantage of the growing recognition of the importance of improving awareness and appreciation of science, where they simply stick with the activities they have inherited from the past, they run the risk that other organisations will assume their mantle, that their own activities will become marginalised and that their continuing existence will be called into question.

5.3. The BA Festival of Science (Annual Meeting)

170 years after its inception the Annual Meeting, now presented as a festival of science, remains the BA's flagship activity. It still takes place in September in an extensive circuit of UK cities, hosted by local universities. Apart from the BA's centenary in 1931 and again in 2000, the festival has always been held outside London - a legacy, perhaps, of the negative view the BA's founders took of the scientific establishment of their day. Today's event is very different from the original, although to describe it as a festival is perhaps misleading. It is certainly very different from the much more recently established Edinburgh International Science Festival, which is described in Chapter 14.

Over 300 presentations on developments in many areas of the natural and social sciences, engineering, mathematics and medicine form the core of the week-long program. In addition, there are public exhibitions, presentations and debates, and 'hands-on' activities for younger children. It is these non-core activities which give the meeting something of the character of a festival. The core participants form a broadly based but attentive audience for science. They include scientists concerned with fields other than their own, teachers, lay people and students from schools and universities. Altogether up to 10,000 might typically attend the festival, with 70%–80% of these being young people.

Its most distinctive feature, however, is the extensive media coverage it receives - far greater than for any other regular science-based event in the UK. The national broadsheet papers regularly provide daily coverage of half to one page, there is some coverage in the middle market tabloids, on national and local radio, and on television as well as in science magazines and other publications. Consequently, the 200 or so journalists who attend are, arguably, the most important audience. Without the coverage they provide, potentially extending the festival's ultimate audience to millions, the future of the event would be threatened. To meet the needs of both the media and others who attend, speakers are asked to talk about developments or issues which are at the forefront of scientific or public interest but to do so in ways that are understandable to an attentive but non-specialist audience.

Looking ahead, the BA is committed to increasing the proportion of sessions that facilitate discussion and debate. However, the media find it difficult to turn such dialogues into 'stories' so that progress has to involve maintaining a balance between what works best for the live audience and what works best for the press.

Annual meetings, in one form or another, remain important events for other associations too, for example in the USA, India and Sri Lanka. Their meetings have not developed in the same way as in the UK. The American Association for the Advancement of Science (AAAS) meeting, like the organisation itself, is a much larger event, with greater international participation, and a more coherent, professional audience, albeit of those whose interests span a wider area than their own specialisation. A crucial difference between the AAAS and other associations derives from its role as the publisher of *Science* magazine, and the fact that subscription to *Science* includes AAAS membership. The importance of *Science* as an international journal helps to ensure that the AAAS has a large membership and the strength that goes with that, and increasingly makes it an international body.

Most associations now arrange some activities for young people at the time of their annual meeting. The AAAS runs a national 'Public Science Day' to coincide with the start of its meeting. The Indian Science Congress Association puts on an extensive exhibit of young people's projects, and until recently ANZAAS in Australia arranged a Youth ANZAAS meeting in parallel to its own congress. In 1998 it was decided to discontinue holding the main ANZAAS Congress because of falling attendance but the Youth ANZAAS activity continues to take place.

5.4. Activities for Young People

The development of science fairs and clubs was an important step in ensuring the continuing relevance of the BA in the 1960s. Today the BA runs a range of activities for young people aged 5-18. Activity days and workshops, arranged in many areas, are designed to excite 8-13 year olds about science. Three award schemes, First and Young Investigators and CREST, aim to develop investigative skills and project work among 5s-8s, 8s-13s and those over 11 respectively. *Visions for the Future* encourages older teenagers to take part in on-line discussions and live events to help them think about developments in science and technology and how they affect their lives. A new scheme in 2001 promotes the development of science communication skills among teenagers.

Such activities have been designed to complement and support the school curriculum rather than to provide a direct means of delivering it. The BA has not been involved in the process of curriculum development. In the USA the situation is different. In 1986, the year in which Halley's comet last passed close to the earth, AAAS launched Project 2061, which has become a major curriculum development initiative and is named for the year in which the comet will next return. The AAAS also runs a range of activities designed to excite children's interest in science and most Associations have a youth program of some sort.

5.5. Science Communication Projects

Through a series of other initiatives, the BA seeks to build links between scientists and the media, to become a focal point for science communication in the UK, and to explore new ideas.

Media fellowships enable young working scientists to spend up to eight weeks on secondment to a newspaper, magazine or broadcasting organisation. The experience helps the scientists to get a better understanding of how the media operate and to improve their communication skills. AlphaGalileo is an internet-based press and expert service for European science. The service, which has similarities to the AAAS EurekAlert, was established because press/promotional culture in many European scientific institutions is different from that commonly found in the USA and needed a service tailored to its needs. Although initiated in the UK, AlphaGalileo now has partners in many other European countries. The BA also provides office facilities for the Association of British Science Writers, the UK's organisation for science journalists. The BA also organises twice-yearly forums for science communicators to exchange ideas and share best practice.

The BA is committed to introducing activities that facilitate dialogue about science-related issues. Public forums are arranged for discussion of one of the topics covered in each issue of its bi-monthly magazine, *Science and Public Affairs*. *SciBArs* are science-based discussions held in local wine bars and similar venues. These tend to attract people from a variety of backgrounds in their 20s and 30s, somewhat younger than those who attend most of the public science communication activities in the UK.

5.6. Looking to the Future

The BA has existed since 1831 by adapting its role in a changing world. Along the way it has come to realise that it does not have a God-given right to exist. It has survived on its collective wits by making sure that it remains true to its historic purposes in ways that are appropriate to the contemporary world. As a result it is recognised today as an organisation with a significant role to play in a field which has ever-increasing importance. To continue playing this role, the BA must face up to new challenges both for science communication and for those involved in it.

Despite the plethora of science communication activities and the multitude of organisations involved, public discussion of issues such as GM crops has been scientifically ill-informed and the number of young people entering science and engineering courses at university has remained static although there has been a significant rise in the student population as a whole. Yet improving 'scientific literacy' and attracting young people into science are the most common objectives of science communication activities. Similarly, in spite of curricular change, young people's experience of science in school seems neither to encourage them into science-based careers nor to equip them for life in a technological society. Far better ways of assessing the impact of current curricular and extra-curricular offerings are needed. There is certainly no room for self-satisfaction within the science communication community; the challenges remain immense.

Surveys show that public interest in science in the UK is high. At the same time the public's trust of some scientists, particularly those working for government or industry, appears to have diminished. Most people get their information about science from the media, whose audiences are far greater than those for science communication activities. For example, more people in the UK read *The Daily Telegraph* each day than participate in all the activities of National Science Week or visit London's Science Museum in a year. When science-related issues become hot news, they tend to be handled not by science journalists but by political correspondents, whose awareness of science is not high and with whom the science community has few contacts. Ensuring that all parts of the media have access to good scientific information, advice and comment is therefore crucial.

The House of Lords *Science and Society* report recognises that public attitudes often spring from deeply rooted values, which must be taken seriously; hence the need for a 'culture of dialogue'. If such a culture is to be developed, we will need many new forms of activity and changed attitudes on the part of scientists. They must recognise that dialogue requires different skills than those involved in one-way communication. It is not just another method of 'getting the message across'; it implies vulnerability, recognition that others might win the argument, that one's own view might be changed. This can be difficult for many scientists, not least because of their understanding of the nature of science. Scientists as well as others need to recognise that public issues have many dimensions other than science.

Despite the increasing numbers of scientists willing to engage with the public, there is scant recognition for their efforts. If research and/or teaching performance are the principal criteria for scientists' career progression and financial security, communicating with the public will remain an optional extra for the committed.

What about the BA itself and similar organisations? Is there still a case in the UK for a broadly-based body that focuses on science communication? Arguably there is. The breadth of its scope and the eclectic character of its membership provide it with appropriate foundations at a time when the public are interested in science but cautious about its role in society and the factors that influence its applications. Unlike the Royal Society, it does not represent the scientific establishment; nor does it wave the flag for specific interests in the way that scientific societies and research councils do.

Consequently, the BA is building for the future. It remains the only nation-wide organisation in the UK, with a comprehensive, multi-disciplinary base and with an open membership, which is wholly dedicated to improving public awareness and appreciation of science. Its goals are to:

> *create firmer foundations* by expanding its membership base and the spread of its activities across the country, and by creating a fit-for-purpose headquarters;
> *foster more open, informed public discussion* of scientific developments and issues through innovation and the development of its existing activities, and by encouraging scientists to increase their appreciation of public perceptions and attitudes and to develop the skills needed for effective discussion and dialogue;
> *strengthen its role* as a key organisation in science communication through its activities and by providing meeting places for those who communicate science to the public, particularly in non-formal settings.

Forming strategic alliances with others is an important part of the BA's strategy. We are working with the Science Museum in London on plans to develop a new building on the museum's site that will house museum and BA staff and provide a natural home for the discussion of science-related issues of public concern. We are exploring with science centres the development of new activities for young people and a UK-wide membership scheme. And we continue to work with existing partners, particularly the Royal Society and COPUS.

6. AN INTERNATIONAL PERSPECTIVE

Science communication, like science itself, is international in scope and there are good reasons to pursue ways of facilitating international co-operation and the exchange of ideas and information. A number of the BA's activities have been inspired by or modelled on those taking place elsewhere. The national science week was inspired in part by the success of such weeks in other countries and the media fellowship scheme was largely modelled on the mass media fellowship program of the AAAS. Many associations, including the BA, already enjoy good bilateral links with each other. Some, especially the AAAS, increasingly operate as international organisations.

It was because of the recognition of the potential for mutual support and inspiration that the BA took the initiative in organising the 1991 meeting in Hong Kong at which IFAAST, the International Federation of Associations for the Advancement of Science and Technology, was inaugurated. Although a few if its member bodies are strong, most of them are relatively weak and it has been difficult to establish an effective means of operation. All the potential members, perhaps especially the smaller ones, stand to benefit from the opportunities that an effective IFAAST could provide. Finding a way of utilising modern communications technology seems likely to provide the cheapest and most effective way to share ideas and experience.

The arrangements for promoting public awareness and appreciation of science in Europe differ extensively between countries. Few have an active organisation dedicated to this purpose like the BA. But all have organisations that run activities, and mechanisms for co-operation do exist, especially through the programs of the European Commission. For example the Commission organises a young scientists contest for school students who have undertaken outstanding project work, and it established a European Science and Technology Week in 1993. To date its impact has not been great but it has provided mechanisms for organisations active in this area to meet. A pan-European organisation, Euroscience, was launched in 1997 with the aim of being an AAAS-like European body. It is too early yet to judge its effectiveness.

7. CONCLUSIONS

Founded nearly 170 years ago with the aim of promoting and advancing science, the BA has focused its work today on improving public awareness and appreciation of science. In this rapidly developing area, it has become a key player in the UK and one to which other bodies look for ideas and leadership. To maintain this position it must be sensitive to the rapidly changing environment in which it works and adapt its operations to new circumstances. While the situation in all countries is different, there is a clear value for such a role in all of them. In some countries, notably the USA, the BA's sister associations continue to fulfil the full range of their original remit; in others the associations might re-shape themselves for this new role. If they are unable or unwilling to do this, the evidence is that other organisations will do so.

The BA, 23 Savile Row, London W1S 2EZ

14. EDINBURGH INTERNATIONAL SCIENCE FESTIVAL

1. INTRODUCTION

The wonderful thing about contemporary science and technology is the rate at which new discoveries are made. This incessant innovation is a precious gift if you are in the business of engaging the public with science and technology. It provides a constant source of new material and new reasons to go back to your audience. Furthermore, these innovations build on an established body of scientific and technological knowledge that is immense. If you look in the right places there is a virtually limitless pool of fascinating and accessible material suitable for public consumption.

In 1989 a new event was created in Edinburgh that set out to access this material through a mainstream cultural happening. It was, and remains, the Edinburgh International Science Festival (EISF). The event is a two-week festival of science and technology involving hundreds of events in 30 venues across Edinburgh. Embracing many styles of presentation including talks, walks, shows, films, workshops and exhibitions, the Festival accesses many areas of science especially the contemporary. As a focused period of intense activity, one of its greatest assets is the feeling of energy and excitement that it generates for the visitors.

From the outset the organisers were determined that the Festival should be a place for discussion and debate; everybody was regarded as a participant, especially those in the audience.

Both the vibrancy and the long-term viability of the Festival have stemmed from the many partnerships it has established with external organisations within Edinburgh and beyond. It works with existing cultural organisations such as museums, visitor attractions, art galleries, theatres and cinemas, as well as many professional and academic bodies. In the early years this approach quickly introduced the Festival to a wide range of audiences and within three years it was attracting tens of thousands of visitors.

More than ten years after the first Festival took place the event has grown and improved. It has proved itself a viable alternative to science centres as a means of reaching a large public audience. Some might say that as it has no need for the large capital outlay that a science centre requires, it may be a preferred model. Despite this there are still fewer than ten large science festivals in the world.

In this chapter I set out to do three things. In the first section I provide an account of the creation of the Festival and what it comprises. Your best opportunity to learn about the Festival is, of course, to visit it.

In the second and longer section I detail many operational aspects of the Festival with the specific intent of helping others who wish to set up a science festival. However obvious these observations may seem with hindsight, many were hard won

S.M. Stocklmayer et al. (eds.), Science Communication in Theory and Practice, 203–217.

and are worth mentioning. Of course, our way of running a science festival is just one of many ways and there is much to be gained by talking to others around the world running similar events.

In the third and final section of the chapter I pose seven critical questions to those wishing to set up a festival. From our experience you will need clear and convincing answers to each of these if your festival is to survive and prosper.

2. THE ORIGINS OF EDINBURGH INTERNATIONAL SCIENCE FESTIVAL (EISF)

City of Edinburgh Council's Department of Economic Development and Estates initiated the EISF in 1989. Its primary motive was to create an event that would boost Edinburgh's tourist numbers over the two week Easter holidays. At that time Edinburgh, the 'Festival City,' was home to six or more substantial arts festivals, most of which took place in August. As a result the economic benefit (in terms of extra pounds spent by visitors in the area) of a successful festival was well understood. Furthermore, as many of the skills and civic procedures to run festivals already existed, starting a new festival was not as daunting as it may have been in a town with no festival experience.

Although the potential benefits to the tourism industry provided the impetus and justification for the City's funding, the ambition to establish a wholly new type of public event that celebrated science and technology in an energetic and accessible way was the creative force driving the venture. This new Festival of Science and Technology was to provide an arena in which the scientific community and the public could mix, exchange views and learn from each other. From the outset the notion of creating events that deliberately broke down the barriers between presenter and audience was regarded as essential and at that time distinctive within the UK. As Ian Wall, one of the principal founders of the Festival, constantly reminds us, everyone at the Festival is a participant irrespective of which side of the lectern or curtain they sit. In this respect the use of the word 'festival' was well chosen even though to many it was a conceptual challenge to link it to 'science'.

2.1 UK and World Context

The foundation of the Festival coincided with a sea change in the UK government's views towards the 'public understanding of science' (PUS). The publicly funded Committee on the Public Understanding of Science (COPUS) was established around the same time to assist groups actively involved in running PUS events. There also appeared to be something of a ground swell of popular interest in science that was either initiated or skilfully echoed by the mass media and publishers. Despite the evident public interest in science, the science centre movement in the UK was struggling to develop; the government's new found enthusiasm for engaging the public in science did not stretch to building science centres. (This happened ten years later in 1999 with one-off funding provided by the Millennium Commission.)

In the way of temporary science and technology 'happenings' there were a number of events taking place around the world. The British Association for the Advancement of Science and the American Association for the Advancement of Science held annual meetings. In India science fairs were happening on a regional and national basis and a number of countries were holding annual science weeks.

Within this context, the Science Festival in Edinburgh had an approach that was quite distinctive. No single attribute is sufficient to set it apart from other events; it is more the combination of factors that created the event. It was a 'festival' not a meeting, conference or centre, this gave it an energy that was higher than you could find elsewhere. It set out to establish a genuine openness and dialogue between the experts and the public. It wanted to be accessible and made use of 30 venues across the city including familiar places such as shopping centres, cafes, book shops, art galleries and cinemas. Finally, by drawing in dozens of organisations to stage events, it had a variety of event types and content that no one else had.

Over the last 12 years a number of other science festivals have taken root around the world, some one-offs and others annual occurrences. Canberra in Australia, Dunedin in New Zealand, Grahamstown in South Africa and Shanghai in The People's Republic of China are well established, regularly occurring events. Within the UK some half dozen science festivals now exist of which three are in Scotland. Conspicuously, there is not to my knowledge a science festival in North America although this seems set to change very soon with plans in place for Pittsburgh, Pennsylvania and Sudbury, Ontario.

2.2. What is the Edinburgh International Science Festival?

Whilst the scale and style of the Festival has changed enormously over the years, the basic formula for the event has remained largely unchanged. Table 1 below shows the Festival's three main audiences, and the specific program of activity organised for each.

A fourth but minor audience is made up of people attending professional meetings associated with the Festival.

The events for the Festival come via two routes.

Events Created by the Festival Team. Around 50% of the events are designed and created by the Festival team. This involves taking the event from the idea stage to delivery and generally paying for all the costs along the way.

Events Proposed by Other Organisations. The balance of the events comes from a range of organisations and individuals. They tend to manage the majority of their event planning and bear most of their own costs. Amongst this group are a number of local organisations that provide large numbers of events (e.g. universities, the botanical garden, the zoo) and often manage venues of their own. This sub-group we call Festival Partners.

Table 1. Edinburgh International Science Festival: audience and programs

Audience	Type of Events	Where	When
School pupils (in schools)	Schools Program: shows, workshops	Touring to schools across Scotland	Jan-Apr
Parents and children	Family Program: interactive workshops shows, exhibitions, walks	Venues in and around Edinburgh	2 weeks in April
Adults & older teenagers	Popular Science Program: talks, walks, tours, exhibitions, films	Venues in and around Edinburgh	2 weeks in April

2.3. The Scale of Activity: Events, Visitors and Economic Impact

Counting the number of visitors to the Festival can be difficult (Table 2). It is easy to count events, ticket stubs and the rotations of a turnstile, but it is very difficult to determine from these data how many different individuals are involved. The table below gives outline visitor figures from a recent Festival showing the number of 'visits' rather than visitors, that is, if you go to two different events you are counted twice.

Although the headline figure is reasonably high one must remember that, as we are allied to some 30 venues, some of these visitors are visiting events without prior knowledge that they are part of the Festival. They may, for example, be visiting a museum or gallery and whilst there they happen upon a Festival event. It is our estimation that as much as half of the total attendance falls into this category leaving around 75,000 visits accounted for by people who have the Festival program in their hand and are actively seeking out the events.

Notice that the Schools Program events take place not as part of the main Festival but before it and in schools.

The economic benefit of the Festival is calculated using an economic model shared by all the Edinburgh festivals. It shows that with an annual turnover of £500-600,000 the worth of the Festival to the region is of the order of £1.8 million. This model accounts for the additional business the Festival generates either directly through its own activity or indirectly through the expenditure of visitors to the Festival.

If you have access to the Internet search on our name and look at the current Festival programmes; these are the best way of getting a real feel for the event.

Table 2. Visitors to the Edinburgh International Science Festival

Program	No. of Different Events	Attendance (visits)
Schools Program	20 (500 performances)	24,000
Family Program	95 (hundreds of performances)	35,000
Popular Science Program	80 Talks 7 Exhibitions 28 Others	10,000 80,000 6,000
Total	230	155,000

3. OPERATIONAL ISSUES

3.1. Introduction

This section deals in some detail with operational issues and has been written with the prospective Festival organiser in mind. It is a view peculiar to our location, markets and the way we are funded and there are without doubt other and better ways of doing things. Many of the points I make may seem minor, even trivial. The difference between success and failure, however, is so often down to the smallest of things.

3.2. Organisational Structure of the EISF

The Science Festival was established as an independent trading company (EISF Ltd) with a board of directors. EISF Ltd in turn is owned by a parent company with charitable status that, until very recently, was controlled by the City Council. In practice, whilst EISF Ltd relies on and takes great comfort from the support of the Council, it is able to behave as an independent business. This confers on it many advantages, the most important of which is that it sets its own agenda. In contrast to many centrally funded organisations or professional bodies, the EISF has no particular professional constituency to promote, no particular PR agenda to pursue. Its constituency is the general public and school children and its aim is to provide them with what it perceives to be the things they most want to see and do. In addition to this independence, it is a slim organisation unencumbered by committees. This gives it an agility to react and change that has contributed greatly to its success.

3.3. Staff and Office

The Science Festival operates with a core full-time staff of seven. These people work from a modest size office that includes a small workshop. 500 m^2 of storage space is kept outside of Edinburgh for large props and equipment.

In addition to this central team the Festival employs some 70 temporary staff during the busiest operational period. Together these people organise and deliver some 50% of the Festival events. In addition to these staff the Festival Partners, such as the museums and universities, make use of their own staff to run their events. In total the maximum number of people engaged in delivering Science Festival events on any one day is around 110–130.

The following list indicates the main areas of responsibility of the staff employed directly by the Festival, whether in permanent or temporary posts.

Permanent Staff

Director. Overall responsibility for the program content, staff and finances
Marketing and Media Manager. Responsible for marketing and media activity
Business Development Officer. Fund-raising, meeting the sponsors needs, business strategy
Event Manager for Popular Science Program. Selecting, co-ordinating and financial management of the events in the Popular Science Programme (Adults Program)
Event Manager for Family and Schools Programs. Selecting, co-ordinating and financial management of the Family and Schools Programs
Science Communicator(s). Creation of new events for Family and Schools Programs
Office Manager. Book-keeping, payroll, office support

Temporary Staff

Technical Manager. Oversees get-ins and get-outs, on-going technical support during the Festival
Eight Get In/Get Out Technicians. Set up and dismantle venues
Financial Advisor. Overview of financial management and planning
Accountant. Monthly review of accounts
Theatrical Trainer & Director. Trains the performing staff delivering science shows
Box Office Manager. Supervises Box Office operation
Venue Manager. Manages the operation of the talks venues
25 Performers. These people are either young trained actors or exuberant scientists
Ten Workshop staff. These people deliver interactive workshops
Three stage and audio-visual. Technicians. Provide technical support for shows and talks
Ten Box Office staff. Sell and take tickets, organise schools tour, provide telephone information
Five 'Front of House' staff.
Two Media assistants. Assist with media enquiries

In a normal year we experience nine months of planning followed by three months of intense operational activity. Over a matter of a few weeks we go from seven staff to seventy and at certain times of the year the working hours are very long. As this can lead to stress, exhaustion and eventually a disenchanted work force, one of the principal planning challenges we face is to spread the workload as evenly as we can across the whole year.

4. PUTTING THE PROGRAM OF EVENTS TOGETHER

4.1. Event Planning Strategies: Is it an Inclusive or Exclusive Festival?

When it was founded the Festival had an inclusive approach to putting its programs together. If people offered events we were likely to accept them, making exceptions only when the content was scientifically flawed or the link to the Festival too tenuous. There was little aversion to people taking 'extreme' views especially if large numbers of the public shared these views. We insisted, however, that we were able to provide some balance by fielding opposing or more moderate views in the same room.

After 2–3 years many organisations realised that we were here to stay, we were being successful at attracting audiences and that it was useful to them to be involved. The combination of this increased interest in the Festival from potential participants and our inclusive approach to planning meant that the number of events at the Festival grew very rapidly from around 80 to 200. This quickly put the Festival on the map.

The disadvantages of this inclusive approach, however, took us a few years to fully appreciate. The problems it brought included the following: the quality of events was variable; some were fantastic and others quite awful; the cost to the Festival of marketing the events was very high as there were so many of them; and the cost effectiveness of many events was very poor as they were being run for too small an audience. Many organisations had low expectations of attracting a large audience.

In 1998 the Festival fundamentally altered its selection process, adopting an exclusive rather than an inclusive approach. Selection criteria were introduced against which all events were judged and potential contributors were asked to submit proposals for us to review. These selection criteria were aimed at optimising our chances of putting together a consistently high quality and popular program in which each event met a minimum level of cost effectiveness. It makes better economic sense to fill 200 high quality events than half fill 400 of mixed quality.

Discussion has continued about whether there is a place for a Science Festival 'Fringe'. This would take an inclusive approach, make no value judgement about content or quality, but provide a focus and marketing tool for those who wanted to get involved.

4.2. What Type of Events are in the Festival?

Talks: The majority of talks involve a single speaker talking for 45 minutes with 15 minutes of questions. A chairperson introduces the speaker and then manages the questions. We look for the very best communicators to talk about popular subjects.

Interactive workshops: These are staged for adults, teenagers and younger children. They cover a wide range of activities that has included cooking chocolate, emergency surgical procedures and electronic assembly. We are interested in 'extraordinary' activities that people would find impossible to do at home or in school.

Science Shows: Performance makes up a large part of our family and schools program. We favour the use of unusual costumes (such as a bumble bee or a 'conehead'), traditional acting skills and unusual environments in which to stage the shows such as a giant set of lungs.

Exhibitions: Many types of exhibitions can be found at the festival ranging from those normally found in interactive science centres to photographic or portrait exhibitions. This reflects the range of audiences that attend the Festival, the range of participating organisations and the breadth of subject matter the Festival encompasses.

Walks and Tours: Visits to unusual places with a scientific theme are popular. A crematorium, the police forensic lab, a nuclear reactor or a disused quarry, they all have their fans.

Film and Theatre: Each year the Festival co-operates with local cinemas to stage a showing of films that link to the Festival content. Similar arrangements with local theatres are also made although much less frequently as the repertoire of relevant plays is currently very small.

4.3. Event Management

Putting together the Schools, Family and Popular Science Programs occupies between three and four people full time for a year. Some of these have science backgrounds, others not. The Festival has a Programme Planning Committee (PPC) that comprises academics that are skilled communicators (many are authors as well as scientists). The members of the PPC provide shape to the program by setting six content strands each year. As we cater for a broad range of interests we do not adopt a single theme for the Festival. PPC members are also well placed to suggest individuals or organisations that might contribute events.

Event planning is at the hub of our existence and from it follows almost everything else we do as an organisation including fund-raising and marketing. Here are the key milestones in the event planning calendar:

18 Months before the Festival. The Program Planning Committee sets six content strands.

June, 10 months before the Festival, event planning starts. Organisations invited to submit proposals.

August. Budgets determining the number of events are set.

September. Schools Program events confirmed. External organisations informed of our decision on their event proposals

October. Copy deadline for Schools Program. First copy deadline for Family and Popular Science programs.

November. Schools Program printed and distributed. Final copy deadline for Family and Popular Science programs.

December. Family and Popular Science Programs compiled, edited and proofed.

January. Family and Popular Science Programs designed and proofed. Schools program activities start

February. Family and Popular Science Programs printed and distributed. Travel, Accommodation, Venue and Technical requirements confirmed with participants.

March. Travel tickets sent out, technical information passed to Technical Manager

April. Science Festival!

To plan events successfully a number of key competencies are essential and as this is core to what we do I have taken time to articulate these. I would urge any prospective Festival organiser to be sure to get all of these characteristics in every member of the event management team.

Attention to Detail and Administrative Competence. Each event has at least eight essential pieces of information related to it that you *must* get right, namely: date, start time, finish time, place, price, capacity, technical requirements and title. Get anyone of them wrong and at the very least you will look incompetent and disappoint a lot of people. At Edinburgh we have some 200 events and some 1200 separate performances of these events; that is 9,600 pieces of essential information which we must get right. Whilst standard forms and computer systems ease the management of these data, I would urge you to let only the most highly organised people anywhere near event planning.

Technical Understanding. It is very probable that the event manager will have to discuss with participants technical specifications that may range from power supplies via video formats to coning off streets to allow lorries to park. Whilst this person does not need to be a technician they do need to have some background knowledge of these issues if they are to collect and then pass on the correct information to those that need it.

Budgeting Skills. At the Science Festival event managers handle their own budgets. It is essential that they are able to do this professionally. Festivals are very open-ended affairs and it is very easy to find ways of spending 50% or even 100% more than you have. Furthermore, there are quite a few risks associated with new Festivals; you are occupying new venues, running dozens of new events with new staff. Failure to anticipate these risks can quickly lead to overspend.

Add to these attributes, calmness, effective communication skills and the ability to say 'no' when you really need to and you have a perfect event manager.

4.4. In-house Production of Events

When the Festival started it was clear that whilst it was easy to get limitless numbers of people to talk at the Festival it was going to be much harder to book good hands-on activities and science shows for the Schools and Family programs. Our reaction to this was to create our own events and some of the very best things that have happened at the Festival have been created via this route. We focus on activities that are lightweight, quick and easy to set up (i.e. 30 minutes to 2 days maximum) and focus on the interaction between the visitor and the performer or demonstrator - that is, shows and workshops but not exhibitions.

Over the years many science shows and workshops have been produced by Festival staff including MadLab, the electronics assembly workshop, theatre style shows such as Imagine That! and Up Yer Hooter! a giant installation involving a 3.5m high nose and lungs into which you climb.

5. MARKETING

In this section I will look at various issues relating to the marketing of the Festival to its key target markets.

5.1. What are we selling?

Whether they are talks, shows, exhibitions, walks or workshops we are in the business of selling tickets to events: our marketing effort is about maximising the number of bums on seats. In doing so we simultaneously maximise the usefulness of the Festival to schools and the public, the revenue from ticket sales and the usefulness of the Festival to sponsors.

Every one of our events we expect to have all of the following characteristics: (i) about science and technology (ii) educational (iii) involving (iv) accessible (v) extraordinary! (vi) and, for schools events, links to the curriculum

All these factors govern our approach to marketing and promoting the Festival.

Table 3. Target market for promotional material for the Festival

Target Market	Decision Maker	Promotional Material	Release Time
Pupils in Schools	Teacher	1^{ary} & 2^{ary} Programs (5,000 copies of each) Poster (5,000 copies)	8 wks before events start. 6 wks before box office opens
Family groups	Parents	Family Program (170,000 copies)	5-6 weeks before events start Box-office opens with program launch
Adults and older teenagers	The indvidals themselves	Popular Science Program (170,000 copies)	as above

5.2. Who are we selling to and with what?

The Festival has three primary target markets and in each there are decision makers that we need to persuade to book events (Table 3).

The Schools Program audience is spread across Scotland with the majority residing in the central belt between Edinburgh and Glasgow. The audience that attends the main Festival is split in the following way: Edinburgh residents, 50%; other parts of Scotland, 35%; other parts of the UK and overseas, 15%. The majority of these people are in the middle-income bracket.

To date it has been hard to justify substantial promotional activity outside of Scotland. This is not because we think people are reluctant to travel to Edinburgh for the Festival but because it is cheaper to promote the event in Scotland and it is easiest for us to attract visitors from our doorstep.

6. DISTRIBUTION OF PROGRAMS, POSTERS AND THE USE OF DIRECT MAIL

Producing good promotional material is only any use if you can get it into the hands of potential visitors. This section looks at how we attempt to do this.

Distribution of Schools Programs and Posters and Selling to Schools. The Schools Program and Poster are sent directly to the majority of schools in Scotland using the education authorities internal mail systems. We also mail directly to teachers that have booked in recent years. The advantages of direct mail are first, you are communicating with someone who already values your events; second, you can personalise the letter, which makes the job of selling more effective; and third, you can draw teachers' attention to specific events.

Distribution of Programs and Posters. There are three channels of distribution for the Family and Popular Science programs and posters; insertion into a Scottish national newspaper (accounting for 50% of our programmes); distribution to public access buildings such as pubs, swimming pools, libraries, cinemas and tourist information offices etc. across Scotland; and direct mail to past bookers and our other databases.

Pricing Policy. It is our aim to charge an entry fee for all events. When the Festival started entry costs were cheaper than those of comparable visitor attractions but in recent years we have chosen to set the prices at levels competitive with similar attractions. Even though this has been done the revenue from ticket sales amounts to only 20% of the Festival's turnover.

Box Office and Information Lines. As with any other visitor attraction or event, making it easy for potential visitors to get the information and tickets they want is essential. To this end we operate a centralised telephone and e-mail request line for programmes and information. We also have a centralised box office service with more than 90% of Festival tickets available from one of two box offices by telephone, post or in person.

Dressing the Town. With a supportive local authority familiar with the needs of Festivals many opportunities present themselves to dress the town. The scope for exploiting these is often limited by resource; however, the Festival regularly makes use of posters, flags, hoardings, banners, shop window displays and branded vehicles to raise its profile.

Market Research. It is very hard to make informed decisions about marketing a Festival if you do not have good information both about your visitors and about the people that stay away. Where people come from, how they heard of the Festival, how they got the program, how happy they are with the events, which papers they read, how much they earn, how old their children are, why do they not come? are all very valuable things to know. Professionally executed market research is the only easy way of collecting this information.

Financing the Festival: Expenditure and Revenue. The Festival costs around £600,000 to stage each year (at 2000 prices). Appendix 1 shows where this money is spent and where the funds to run the Festival come from. These figures do not include the costs incurred by the Festival partners who are working with us to stage events. We estimate that in an average year they are spending a further £200,000, most of this in staff time. Within the operational budgets of the Schools and Family Programmes some 70% of the expenditure is spent on people and of this some 15% is spent on training them. This reflects the Festival's distinct preference for activities that are orientated around interactions with our staff rather than objects such as exhibitions. Even after 11 years of operation cash sponsorship from the private sector accounts for less than 20% of company turnover. One final observations is worth making. Festivals are a temporary happening and for this reason expenditure on capital items such as exhibitions and technical equipment is generally less cost effective than hiring.

Fund-raising for the Science Festival. From the outset funding the Science Festival has been difficult and at times extremely difficult. A number of specific factors have contributed to this. In the early years, the event was little known; it had no track record, and was of variable quality. It has taken some time to develop the 'products' we are able to sell to potential sponsors and the Festival does not commit adequate resource to fund-raising. At a national level, the UK has no substantial source of public funds available for this type of activity. The total UK government's revenue expenditure on public understanding of science activities is currently below £3 million per year. Most of this is earmarked for national institutions and the remaining money is distributed in small packets of typically less than £10,000. This picture is in stark contrast to the areas of Arts, Sport and Heritage in which many hundreds of millions of revenue pounds are expended annually. The absence of a wealthy national public funding agency is a substantial handicap for organisations like the Science Festival and massively frustrates the growth of this area of cultural activity in the UK. We have, however, been fortunate in Edinburgh as the City Council has remained a solid moral and financial supporter and in doing so has underpinned our survival. In this respect I believe the City Council has demonstrated exceptional foresight in attempting to establish a cultural mix that includes science. The starting point for a successful fund-raising bid is understanding why a potential sponsor might want to be involved with you, that is, to identify what you have that they want. For the Science Festival there are quite a number of reasons why organisations fund us, summarised in Table 4. Matching the needs of the sponsor with what you can genuinely deliver is the key to securing support. In our case this has often involved the modification of many aspects of the Festival to make it easier to secure this support. Having a substantial level of marketing activity, being able to deliver high numbers of visitors in most areas and being reasonably confident of the quality of the activities across the Festival are our principle assets.

Media Coverage. The Science Festival has three quite distinct motives for attracting media coverage. The first is to boost visitor numbers. Media coverage of Festival events is a cost effective promotional tool that leads to additional visits in the same year. This applies to the Schools and the Family and Popular Science programs. The majority of this type of coverage is localised to Scotland, which is where 85% of the Festival visitors come from. The second is to promote world class scientific and technological innovation taking place in Scotland. The promotion of Scottish innovation, one of the Festivals' stated aims, involves media coverage that extends throughout the UK and beyond. This type of coverage does little to stimulate additional visits to the Festival in the same year; however, it must have a beneficial role in promoting the Festival throughout the UK in the longer term. This coverage is achieved through the two day Scottish Science Press Conference. It is staged specifically for the UK and international media and focuses on the research taking place in Scottish laboratories. Third, and finally, there is the need to portray our sponsors positively. For many sponsors positive brand association with the Festival is partly why they have become involved. Being successful in promoting this association through the mass media is something they value highly. Being successful in achieving this coverage requires careful planning involving appropriate photo calls, naming of events etc.

Table 4. Interests of Potential Sponsors

Motivation for involvement	Who likes to support?
Educational role with the public	Public bodies, trusts
Educational role with school pupils	Public bodies, trusts
Profile raising of innovation in Scotland	Public bodies
Generating a positive association between 'Edinburgh' and 'Science'	City Council
Economic benefit to the city and region	City Council and local enterprise company
Brand Awareness	Private firms (marketing budgets) and Public bodies
Community involvement/Good citizenship	Private firms (community budgets)
Corporate entertainment	Private firms

7. SO YOU WANT TO RUN A SCIENCE FESTIVAL?

If you are thinking of running a science festival take full advantage of the body of experience that exists around the world; festival folk are rarely proprietorial about what they have learnt. Your new festival will, of course, have its very own reasons for existing and its own operational constraints but if you want a festival that really flies you could, in my view, do worse than to have sound answers the following questions.

Exactly What Are You Trying to Achieve? What do you expect your primary contribution to the world to be? Is it the raising of scientific literacy amongst the general public? Is it to educate and inspire school children? Is it to feed the media stories? Is it to generate economic benefit for your local area? Is it to promote your location or your organisation or your sponsors? Is it to improve the communication skills of scientists? Most likely it will be a mix of these and other reasons but select, prioritise and be clear about how much resource (i.e. people and money) you will commit to each objective.

Do You Have a Stable and Sustainable Funding Base? Science festivals are not profitable. So how are you going to pay for yours, who are your sponsors and are there enough of them for you to still be in business if one or more walked away?

Can You Run Your Festival as a Business Rather Than an Institution? From my observation, it has been EISF's ability to operate as a business rather than an institution that has been one of the keys to its success. It is unencumbered by committees, institutional politics and deadwood. Will you have the freedom to be a business?

What Range of Production Skills Do You Want to Have? Do you want to book others to deliver all of your events or do you want to have the in-house ability to produce events of your own?

Can You Recruit and Keep Talented, Dedicated Staff? Festival staff members have to be adaptable, resourceful, infinitely calm and able to work very long hours. Can you get and keep people like this?

Who Will Your Partners Be? Festivals draw people and organisations together. Who will these people be at your festival, are they the right sort of people and can you establish a good working relationship with them? You might also want to consider the nature of your relationship with your partners; what does each side think they are doing for the other and how much is this valued?

Can You Keep Control of the Festival? If you run your festival well, you will be the best person to know what should be in your program. Can you exert a sufficient level of control over your partners to ensure you do have a fantastic event?

Festivals are very dynamic, energetic and fluid types of event. Their flexibility makes them a natural home for people who are prepared to innovate and to dare to do what others won't. So be adventurous! We are very willing to share our experiences with others. Please contact us or, better still, visit the Festival!

Edinburgh International Science Festival
esf@scifest.demon.co.uk; www.edinburghfestivals.co.uk/science

Appendix 1. Edinburgh International Festival: Expenditure and Revenue
(All figures are £ sterling at 2000 value)

Overheads	
Permanent Staff Salaries, Pensions, Taxes	161,500
General Business Overheads	55,000
Company Contingency	15,000
Total Overheads	**231,500**
Schools Program Costs	
Events	70,000
Marketing	7,500
Box Office	7,000
Media	1,000
Total Schools Program Costs	**85,500**
Popular Science Program	
Events	23,000
Marketing	20,000
Box Office	7,000
Media	3,000
Total Popular Science Program	**53,000**
Family Program	
Events	71,000
Marketing	19,000
Box Office	7,000
Media	1,000
Total Family Program	**98,000**
One Off Projects	**75,000**
Development Fund	**57,000**
TOTAL EXPENDITURE	**600,000**

REVENUE	**Net of VAT**
City of Edinburgh Council	125,000
Public Sector Grants & Sponsorship	235,000
Private Sector Sponsorship	103,000
Schools Ticket Sales	53,000
Popular Science Programme Ticket Sales	20,000
Family Programme Ticket Sales	42,000
Franchises	2,000
Partners contributions	20,000
TOTAL FUNDS RECEIVED	**600,000**
Advertising sales in programmes	**20,000**

M.M. GORE

15. THE QUESTACON STORY

1. BEGINNINGS

I might never have got into the interactive science centre industry if it had not been for the infamous American Senator Joe McCarthy.

In the early 1950's the Americans - or at least those on Capital Hill in Washington - had a preoccupation with looking for 'reds under the beds'. Senator Joe McCarthy ran his own anti-Communist witch-hunt, that resulted in many innocent and harmless people having their lives and careers turned upside down. One such person was Frank Oppenheimer.

Frank was the brother of atomic physicist J. Robert Oppenheimer. He was teaching physics at the University of Minnesota when he was caught up in the McCarthy panic. While Frank was a student at Berkeley University, he had flirted with the communist movement on campus. This came back to haunt him in the 1950's when the University of Minnesota fired him for having been cited for un-American activities.

Faced with a hiatus in his career, he established the Exploratorium interactive science centre in San Francisco. It immediately became a classic pioneering project on the international scene. Interactive science centres began to blossom and flourish everywhere. Their spread across the world really dates from the establishment of the Exploratorium, even though interactive exhibits had been introduced into museums like the Munich Deutsches Museum, the Palais de la Découverte in Paris and the London Science Museum as early as the 1930's.

I discovered the Exploratorium in 1975. It had a very marked effect on me, so much so that I decided to build in Canberra a small prototype of what I had seen in San Francisco. At that time I was teaching Physics at the Australian National University (ANU).

2. THE GENESIS OF QUESTACON

Interactive science centres have proved to be very popular and powerful way of promoting science to the public all around the world. They differ from other methods of interesting the public in science in that they employ exhibits with which the visitors can experiment. The term 'hands on' is commonly used to describe the interactive devices that are used by visitors. 'Interactive', however, is better because the activity in which the visitors are engaged does not always call for the use of the hands. Frequently one or more of the other senses are brought into use.

Stephen Pizzey (personal communication) recently said that interactive science centres and the English Victorian music halls have a lot in common. The music halls became very popular because people with no formal knowledge of music, and no training in singing could really enjoy themselves. In visits to interactive

219

S.M. Stocklmayer et al. (eds.), Science Communication in Theory and Practice, 219–235.
© 2001 Kluwer Academic Publishers. Printed in the Netherlands.

science centres, people with no formal training and little knowledge of science can enjoy themselves and perhaps acquire a little learning.

In the late 1970's the enrolments of students in science at the Australian Universities had begun to dwindle alarmingly. This, I suspect, was one reason why the ANU Council gave me permission to establish, in Canberra, an interactive science centre under the auspices of the University. On all counts, it was a rather unconventional pursuit for an academic but very soon I found I was not alone. Others followed suit and many science centres in Australasia were subsequently launched from universities. In Australia, it happened in Newcastle, Wollongong, Bendigo, Brisbane and Perth. In New Zealand the universities in Palmerston North and Christchurch both played a role as midwife to their interactive science centres. When Questacon opened in 1980, however, it was the first interactive science centre in the southern hemisphere.

It was really about being in the right place at the right time There were four main tasks that needed to be addressed at the beginning of the project. I had to find money, staff, exhibits and a home. History has shown that emerging science centres obtain their initial funding from any number of sources. In the case of Questacon it came from a Federal Government innovations program. This program provided small amounts of seed funding to people in Australian primary, secondary and tertiary teaching institutions who were judged to have worthwhile ideas that might somehow link into the formal teaching system. The grant that I won was $50,000 and it achieved little more than to make the project respectable. Much of the subsequent finance was obtained through the goodwill of various sections of the Canberra community.

2.1. Finding a Home: The Ainslie Infants School

The accommodation for Questacon was provided by the Canberra Education Authority. A school not far from the University campus became vacant as a result of dwindling enrolments due to an aging local community. The education bureaucrats recognised that an interactive science centre, initially to cater to the needs of the local school children, would be a valuable community asset. Questacon thus started life in a school building that was perfect for the purpose. The space consisted of six large class-rooms all grouped around the original assembly hall. They were to house the permanent exhibits. In addition there was an office, toilets and what had once been the school library soon became the venue for science shows.

The *really* attractive thing about the building was the fact that it was leased to the ANU at no cost. The terms of the Questacon's tenancy were that all building maintenance costs were covered by the Education Authority together with the cost of heating the building in winter. A telephone was also available at no cost for unlimited outgoing local calls. It was not possible to make long distance calls but it was possible to receive incoming ones. This facility became very important as Questacon grew and bookings from school groups began to come in from further and further afield.

2.2. The Early Operation

I had no experience of running a science centre so I was naturally very cautious at the beginning.   There were no full time staff members, neither clerical nor technical, so I decided not to open the centre seven days a week.   I knew that the resources available to Questacon at that time would be quite inadequate for sustained operation.   I well remember worrying about how many people would turn up if Questacon were indeed opened on a full time basis.   I did not have sufficient Explainers to run the place over such an extended period and I instinctively knew that the interactive exhibits would not stand up to a large volume of traffic. Although I had wonderful support from the workshop in the ANU Physics Department where I was based, the first call on their skills was naturally from its teaching and research areas. I carried out exhibit maintenance myself, together with a small, dedicated team of Explainers who had the necessary practical background.

At first, Questacon did not open to the general public.   Instead, I decided to run a booking system for schools, a policy that was consistent with the terms of the original grant. I divided each operating day into two sessions; one before and one immediately after lunch. Each session was of 90 minutes during which one group of 45 students was let loose in Questacon. The size of the group was dictated by the capacity of the bus that brought the students.

The main purpose of having Explainers on the floor of a science centre is to put a human face on the science. I arbitrarily decided that in each session I would have a ratio of five students to every Explainer.   My choice of explainer/student ratio was influenced by a visit I paid at the time to an interactive exhibition organised by a major telecommunications company.   The exhibition was in the foyer of the company's building and the only employee in the area was a receptionist.   I asked her about the exhibition and she told me that although it was very popular she had observed that it had a big problem. She told me that teachers brought in groups of school children and then left them to it while they themselves went off for coffee. The result was that some of the unsupervised students created havoc.   From this useful piece of intelligence I made it a Questacon policy that one or more teachers would always be required to stay in the building with their charges and that they, not the Explainers, would be required to keep their students under control.

For the first few months Questacon operated just two days a week, that is, for four sessions of 90 minutes.   During that time it entertained some two hundred students. Well-defined visiting hours meant that I could tell Explainers exactly when I needed their services. We made a modest charge for each student, one dollar each with no charge for the accompanying teachers.   Nor was there any charge for parents who chose to tag along. We always took the teacher's word for it, when they told us how many students there were in their group.   Sometimes there would be a student from a poor family. The teachers usually knew who could not pay and why, and for this reason we never counted heads. Explainers were paid a small fee per session, leaving a small surplus for the Centre.

Making Questacon bright and colourful took an awful lot of my time–and that of many others–in those early days.   I recall that some colleagues and I cut and painted sign boards fire engine red while another cut out literally hundreds of polystyrene letters.   We hung the signs, proclaiming such things as LIGHT, ELECTRICITY, and SOUND, from the rafters.

3. QUESTACON SPREADS ITS WINGS

Eventually, after the first year of operation, I became sufficiently confident to open Questacon to the general public on occasional weekends. This was a big step at the time for I had no idea what would happen. First, I opened from 2pm to 5pm on Sunday afternoon.   The response was so good that we began to open all day Sunday, and eventually all day Saturday as well. An essential part of my policy in those days was to take one small step at a time and observe what happened. This is a tactic I earnestly recommend to anyone just starting up.

Questacon proved such a great success that, by January 1982, it had become one of Canberra's leading tourist attractions.   At first, when it was opened to the public, there was no admission charge; I simply placed a box in the entrance foyer with the written suggestion that visitors might like to make a contribution to the running of the centre. I soon discovered the flaw in that tactic and it was not long before I began to charge admission!

January is the height of summer in the southern hemisphere, with temperatures occasionally reaching 40°C. The combined effects of an Australian summer and the popularity of Questacon rendered the main hall almost uninhabitable. It regularly got so crowded that it was difficult to move around and use the interactive exhibits and the heat made a visit very tiring for everyone except those with phenomenal temperature tolerance.

It was then that I devised what the military call a diversionary tactic.

3.1 Introducing Science Shows

I decided that it would be nice to provide visitors with a little entertainment, which they could enjoy while sitting down and relaxing. This same tactic is often used at world expositions, outside theatres and in Disneyland where there are long, slow moving queues. To keep the visitors happy, entertaining diversions are staged. This was how science demonstration shows made their appearance at Questacon.   Equipment was borrowed from the Physics Department on Friday afternoon, used in science shows over the weekend and returned on Monday morning. Eventually I was doing more face-to-face "teaching' at the weekends than with my undergraduate classes during the week.

The first show I developed started with a demonstration of how one could cause a fluorescent tube to light up by bringing a Tesla coil close to it. I followed this by dropping a coin and a feather in a transparent tube from which most of the air had been removed. They fell, of course, at the same rate. The reaction to this particular demonstration taught me that there are many phenomena that people, including scientists, know about but have never seen. Most of the members of my Physics Department admitted that they had never actually seen this classic 'guinea and feather' demonstration performed.

Shows became an important part of the list of attractions at Questacon. They gave us the opportunity to show the public things that did not lend themselves to the interactive treatment for various reasons, of which danger is one. The Liquid Nitrogen Show, for example, is an extremely popular science demonstration. Unfortunately, liquid nitrogen is not a substance on which the public can be let loose because misuse may result in serious injury. Science shows thus began as a way of giving the visitors a rest but they soon expanded into a major Questacon offering.

4. THE QUESTACON EXPLAINERS

The next problem was where to find staff. Fortunately, Canberra has a large number of scientific research organisations. I contacted them to get the names of scientists and technicians who had recently retired. I also got in touch with professional bodies like the Institution of Engineers, the Australian Institute of Physics and the Royal Australian Institute of Chemists. I called a meeting one evening at the University and invited the recent retirees to come and hear about the project. That evening there assembled a small, but what turned out to be a very keen and committed group of explainers. Naturally, none of them had ever been involved in the development of a science centre. We were, all of us, rank amateurs.

They soon organised themselves into a roster. One of the most difficult aspects of running an explainer system is operating a roster. It is a big task and now occupies one person full time at Questacon, which has about 250 Explainers. The volume of visitors in the centre varies from day to day and even during the course of a day. Because of this the number of Explainers needed on the floor at any given time changes. It is desirable to tailor the number of Explainers present at any time to the number of visitors. This can only be done successfully if there are a lot of pre-booked groups. Many science centres have established teams of explainers. Sometimes they are paid small honoraria for their services and sometimes they are completely voluntary. The advice that I was given at the start of the Questacon Project was that by providing a small honorarium we would get a much greater commitment.

On a less happy note, but of great importance, most science centres, and certainly all Australian ones run security checks on prospective explainers to ascertain if there is any record of child molestation.

I believe the Questacon project was successful because of the strong commitment on the part of everyone involved. The same sort of thing has occurred in other science centres around the world. It is not unusual for science centres to begin in a very modest way in humble quarters with the help of a bunch of committed people who are passionate about what they are doing. No project is ever likely to succeed unless passion is one of the ingredients.

This truism has been demonstrated in science centres over and over again around the world. John Beetlestone developed the Techniquest initiative in Wales. From his position of Professor of Science Education at the University of Cardiff, he recruited a team of enthusiasts and opened his science centre in an old shop in the middle of the city. It was eventually transferred to a purpose built home in the

Cardiff Docks development. Professor Tim Roberts at the University of Newcastle in Australia went down exactly the same path when he established Supa Nova. His fledging science centre started life with a couple of dozen volunteers and opened on the top floor of an old department store. In Bristol, England, the guiding light was Professor Richard Gregory and the Exploratory got off the ground due to the commitment of a few dedicated people. This eventually metamorphosed into the new millennium centre @Bristol. In Vancouver, BC, in the early 1980's Carol Tulk recruited a young and enthusiastic team which established the city's first interactive science centre in an old department store in the middle of town.

In the beginning I only saw the Explainers in terms of how they could help Questacon. It soon became apparent, however, that the program was a fine example of symbiosis. The older Explainers found that being involved with Questacon was very satisfying. Many of them were professional people and even after retirement had a lot to offer to the community and were eager to do so. Questacon provided them with a way to go on to using their skills and knowledge in a project that was clearly of benefit to the community; conversely they are of great benefit to Questacon because they give the science a human face.

Once they had learned how to deal with the public, they began to generate tales and snippets of information that linked the science on display with what the visitors already knew. The art of the Explainer is to develop ways of showing visitors how science is relevant to them.

Quite early on in the development of Questacon, I discovered that it is not essential for Explainers to have a science background. Indeed, there are some who argue that scientists make poor explainers because they have a tendency to try and tell visitors everything. Such explainers have been seen to button hole a visitor and literally deliver a one-on-one lecture!

Once the visitor has explored the exhibit, the Explainer assumes the very important role of showing how the science is exploited in everyday life. This information can be incorporated on graphics panels adjacent to the exhibit but having a person explain is much more powerful. Although this is perhaps the most important role it is not the only one. The very presence of Explainers on the floor of the science centre tends to curb the more exuberant younger visitors whose enthusiastic use of exhibits can markedly increase the maintenance load. This is not to suggest that Explainers are employed as minders, it is simply that having a uniformed person in attendance is sufficient to rein in the over-enthusiastic.

The Scitech Discovery Centre in Perth (Western Australia) has two distinct types of people who deal with the visiting public. One set are volunteers who receive no payment and spend time on the floor talking to the visitors about the exhibits. The second group is paid and its members have a range of roles depending on their skills. They can be involved in admission sales, giving science shows, busking, running birthday parties, and many other day to day science centre activities.

Questacon has always been at pains to provide training for its Explainers. Training deals both with the science associated with the exhibitions and training in the actual mechanics of dealing with the visitors. Each year a series of lectures is held to provide them – young and old alike – with an introduction to the science that is on show in the galleries. It has proved very popular, especially with the students in that they have been able to gain credits in their high school courses.

5. THE EARLY EXHIBITS

The biggest challenge of all was how to build the interactive exhibits. Here again I turned to the Canberra community. Canberra is a small city by world standards; only 300,000 people. A surprisingly high proportion, however, work in institutions which are concerned with science and engineering. Canberra is home to the Australian National University, the University of Canberra, the Australian Defence Force Academy, the Canberra Institute of Technical and Further Education and the Commonwealth Scientific and Industrial Research Organisation. I talked to the heads of research institutions and asked if each would build an exhibit to my design. I explained that there was no set deadline for the completion of a particular exhibit but that, in every workshop, there were slow periods and I asked that my work be slotted into these quieter times. I also suggested that they donate their time to the project, as it was being run under the auspices of the ANU. This University stewardship gave the project the respectability it needed and I not only got workshop time for nothing but, in most cases, there was no charge made for materials.

Where an exhibit needed an expensive component I appealed to various companies for help in donating the required equipment. Many of the Canberra firms and businesses helped in this way and it had the effect of turning Questacon into a thriving grass-roots community project. Twenty years on there are still many people in Canberra who have a real sense of ownership of Questacon.

Eventually I succeeded in persuading some twenty-five workshops to build exhibits and one by one they arrived. In the beginning, so as not to overburden the workshops, I kept the exhibits very simple. The basic criterion was that they should cost little or nothing to construct and maintain. This became known as my 'broomstick' philosophy of exhibit design because one was used for an early exhibit. Such items cost very little, can be obtained from any hardware store and if broken or lost can be replaced easily and cheaply, without the help of the technical staff that in any case Questacon did not have!

In the early days I was rather worried about the simplicity of many of the exhibits. I had the feeling that visitors would not take Questacon seriously if the place were not filled with state of the art, cutting edge, high-tech displays. But I need not have worried for it soon became obvious that the two things that the visitors loved about the place were the Explainers and the simplicity of the interactive devices. Far from deterring visitors, the place was crowded every weekend and the public took to Questacon so enthusiastically that soon it was inundated with visitors from much further afield.

I knew we were on the right track, when one day an entry appeared in the Questacon visitor's book; a lady had written

> I have had my three children in Questacon for three hours and during that time they have not once asked once for food or drink.

One of the most important aspects of an interactive science centre is to have ways of repairing the exhibits. Because maintenance is a very real problem I restricted the number of visitors at any given time, which is why we began by having only small, organised groups of school children. Such visitors could easily be regulated, but to impose this restriction on the general public would have been very difficult. For the first four years, maintenance was carried out mostly by technical staff of the Physics Department, where I was still teaching full time. I would also spend time in the evenings and at the weekends, when Questacon was closed, repairing and cleaning the exhibits, in which task I was aided by many of the Explainers.

The Physics Department also provided another very precious input to the project. One of the secretarial staff of the department was given permission to help me with the administration of Questacon. In the days when Questacon was a part of ANU, Marilyn Miklos typed all the official begging letters, the countless articles promoting the Centre and handled all the correspondence with the hundreds of schools who wished to make bookings. All this work was provided at no charge to Questacon.

Even though it is nearly twenty years since Questacon opened I can still see that half empty main hall, and I remember wondering if anyone would ever take the project seriously. One of the early battles that took place between me and the city fathers was getting Questacon finger posts erected on some of the main streets nearby. Getting nowhere with the bureaucracy, I had the signs made and put them up around town myself. It took the bureaucrats a couple of months to focus on them. They responded by requesting that some, but not all of the rogue signs be removed. This experience was a great vindication of my philosophy, 'Don't ask permission, ask forgiveness'.

Exhibits came from the most unexpected sources. There was one exhibit which came from the school next door. The Principal showed me an elliptical horn that had been part of a loudspeaker assembly used for addressing the students in the playground. Apparently it had fallen off the wall several years earlier, and in spite of many requests to the authorities it has never been replaced. This defunct device was transformed, by the addition of a couple of large steel ball bearings, into the *Gravity Well*.

Those early days of Questacon were very special. The education authority had agreed to take care of the maintenance of the building but I had great difficulty in persuading them that the roof leaked during heavy rain. Then one day Questacon had a royal visitor. The Prince of Wales arrived in a terrible downpour and half the Explainers were on their hands and knees mopping a up couple of inches of water. The incident made television and the maintenance teams were around to fix the problem the next day.

The Australian physicist Sir Marc Oliphant was a great supporter and a frequent visitor to Questacon. He once visited the Centre in company with another eminent physicist, Sir Leonard Huxley. Huxley and Oliphant were like a couple of kids, trying everything and chuckling to each other over the various exhibits. At one stage they stood in front of the two parabolic dishes in the main hall, Oliphant at one end and Huxley at the other. They conversed with each other by speaking into each dish at the focal point. When they came back together again Oliphant said to Huxley, 'You know, I've known about the focusing properties of the parabola all my life, but that's the first time I've ever actually experimented with the effect myself!'

People in all walks of life love experimenting; they enjoy trying things out for themselves. From early childhood everyone becomes involved in experimenting with his or her environment. It is a craving that is innate in all of us but is often suppressed. The interactive science centre promotes an atmosphere that is conducive to messing about. It strips away inhibitions and people become enchanted with what they are doing. I must credit the use of 'enchanted' in this context to my friend Emeritus Professor John Beetlestone, a Fellow of the Royal Institution and the Founder of the Techniquest Science Centre in Cardiff, Old South Wales.

6. DESIGNING AND BUILDING INTERACTIVE EXHIBITS

In the years leading up to the establishment of Questacon as a national science centre in 1988, the skills of the exhibit design team increased dramatically. Today Questacon has evolved into a national institution, and its design team is responsible for all the exciting, and in many cases, innovative interactive exhibits. The experiences that the team has gained since those early days have caused them to formulate a number of concepts which they hold central to good design.

Do not challenge the public to tests of strength. The public will accept the challenge with enthusiasm. Electrical generators are particularly prone to this problem - a good example is the bicycle generator. The visitor is invited to sit on a bicycle linked to a generator and pedal as hard as they can to achieve some goal, such as powering a radio or lighting an electric bulb. The aim is usually to have the visitor get a feeling for what 40 watts represents in terms of their own energy output. Science centres all over the world have found that such exhibits need constant attention and repair. After six months operation one science centre had a large crate full of broken bicycle components to testify to the fact that the visitors had indeed accepted the challenge to pedal as hard as they could.

In its early days, Questacon had a large Wimshurst Machine, constructed from the original drawings by the great man himself. The visitors turned the twin counter-rotating glass plates by a simple hand crank mechanism. The younger visitors would put their backs into the task with a ready will, but all the while would be staring down at the floor. Many never saw the electrical discharges that arced in the cabinet because they were far too busy, with their heads down, cranking the driving mechanism!

One exhibit - one concept. Keep it simple.. Don't try to incorporate too many concepts in any one exhibit. The temptation is always there to hang a few more ornaments on the Christmas tree. Rather than making the concepts clearer, they become confused in the mind of the observer. The more complicated the exhibit the more difficult it is to keep it running and a broken exhibit is a disincentive for visitors. And if it is too complicated it may well not achieve its aim. In any case, simplicity reduces the amount of maintenance required

Avoid labour intensive exhibits.   These are the exhibits that require staff to replenish materials regularly.   As mentioned above, visitors are annoyed and frustrated when they come across an exhibit which they can't use. This is one of the main problems with chemical interactive exhibits in which reagents need to be regularly replaced.

If an exhibit is off the floor more often than it is on, scrap it! One problem that often occurs is that the designer/constructor tries very hard to save that which is doomed, and spends an inappropriate length of time trying to adjust, repair or modify.   Questacon once had an exhibit that used dried peas to enable visitors to discover something about large numbers. Within 30 minutes of it being placed in the display gallery it was scrapped - never to return.   The younger visitors preferred to throw the peas all over the floor rather than use them for the purpose for which they were intended.   The designer believed it was a great exhibit but had not allowed for human nature. I must confess - I was the designer!

Keep up with the maintenance.   Avoid having too many out of order signs decorating the galleries. If possible a defunct interactive exhibit should be removed from the exhibition gallery immediately.   This of course means that exhibits must be designed so as to be easily moved.

Make it colourful.   There is a large body of opinion that believes that interactive exhibitions should be colourful as it makes them far more enticing.   Sombre surroundings do not put visitors in a happy frame of mind and the object of an interactive science centre is to make visitors more relaxed with the science around them.

Don'tuse too many words in the instructions. Only a few visitors will spend time reading long winded instructions and explanations. The Explainers can and should be sources of additional information when requested. Remember, one of the great temptations to resist is telling everyone everything about the exhibit.

People come in different sizes. It is often useful to have two identical versions of an exhibit, one for big people and one for small people. Otherwise small people will stand on things that you do not want to them to stand on, in order to operate big things.

Don't forget people with problems. Many interactive science centres around the world now pay particular attention designing and siting exhibits so that they can be comfortably used by disabled people. There are places where the graphics have been duplicated in Braille.

Adopt the broomstick philosophy. If you can construct an exhibit out of something like a broomstick, then if it is lost or if it breaks it is easy and inexpensive to replace. More to the point, it is not necessary to have a skilled technician to carry out the replacement.   Of course, this is a difficult goal to achieve, but it does provide a benchmark in exhibit construction.

Be aware of the potential for danger in an exhibit.   Exhibits must be carefully vetted for aspects of danger.   In the end just about anything that is used as an interactive exhibit can be misused - and in doing so injury can result.   Often, the best one can achieve is to exclude the glaring examples of potential danger, such as sharp edges or accessible electronic components.

6.1 Things That Didn't Work and Some That Did!

My personal design attempts sometimes ended in disaster.   The bicycle generator taught me that it was not a good idea to challenge the strength of the visitors, who will always take up such a challenge with gusto.   The giant Archimedean screw never worked and stood in the main hall as a mute testimony to my defeat.   Another one that we spent many a weekend fixing was the series of self-starting syphons.   First the internal parts of the pump rusted and the rust ran into all the containers staining the water red.   Next we found that the rust was not the only thing that discoloured the water, there were also the algae to contend with and that took a lot more fixing.   I lost count of the number of times that we took the exhibit to bits, cleaned it, and then reassembled it.   Then there were the problems with the pump and the evaporation of the water.

A telecommunications exhibit that I designed never worked well.   It was supposed to explain geostationary satellites.   I think our visitors went away with very confused ideas about the concept.   A Telecom exhibit demonstrated polarisation of microwave signals.   It played music continuously and it nearly drove explainers mad.   But it did the trick of attracting visitors attention - especially the VIP's who were often impressed by what the exhibit looked like, rather than what it did and how it did it.

There was a fighter bomber in the front foyer.   When the RAAF mechanics came to refurbish it after four years of loving use by thousands of kids, they said that they had seen several similar aircraft after they had crashed and none looked in as bad a condition as the one we had.

I had a personal hand in quite a few of those early exhibits.   When we bought my daughter a new piano, I took her old one to pieces and one of the early explainers built a stand for it.   That exhibit is still going very well, I believe, in the Wollongong Science Centre.   When we moved into the new quarters I donated quite a few of our old exhibits to Wollongong and many of them are still going strong ten years on.

7. SCIENCE THEATRE

Whatever science centre you visit, anywhere in the world, you will find that it presents science demonstration shows, usually about twenty minutes long and covering a multitude of subjects.   Their origins go back a long way - probably back to the beginnings of the Royal Institution at the turn of the eighteenth century.   The Royal Institution, under the leadership of people like Benjamin Thompson, (Count Rumford), Humphry Davy and Michael Faraday introduced a policy of offering popular lectures to the public.

Faraday's lectures are legendary.   He believed in making science accessible and entertaining for the lay public.   He maintained that, in order to hold an audience's attention, it was necessary to 'strew the lecture path with roses'. It is a philosophy that is not espoused by very many scientists even today, but out of these lectures developed the famous Christmas Lectures that are broadcast to millions of television viewers all around the world.   At Questacon, to advertise a presentation as a 'science lecture' was a good way to kill it; the public stayed away in droves. The term 'science theatre' was coined instead.   While 'lecture' has unfriendly connotations, 'theatre' has a much greater pull for the public.

Science centres have many different reasons for mounting science shows.   Apart from the pragmatic one of providing an opportunity to sit down, they offer the opportunity of reinforcing the scientific ideas that are illustrated by exhibits on the floor. They can also present material that is not available in the galleries and if done well they often show things that the visitors can easily attempt at home. Frequently, they are specifically designed to complement a particular exhibition. When Questacon was having a dinosaur exhibition, for example, shows were presented each day about their biology.

Another useful feature of the science show is that it is able to deal with subjects that are difficult to cover in the interactive exhibit galleries or are very labour intensive. With botany there is the tyranny of time. Things take a while to happen and therefore it is generally a subject that does not lend itself to the interactive philosophy.   With zoology the problem is that animals need constant attention, generating costs that few science centres can afford.   Science shows also permit the demonstration of potentially dangerous things to the public.   Chemistry in the display galleries is dangerous, in that the younger visitors can ingest harmful substances. The dangers of chemistry can be eradicated if the chemistry is part of a science show.

Science shows are also wonderful training grounds for staff who need experience in public presentations.   It was this aspect that led Questacon and the ANU to establish a science communication course within the Faculty of Science (see The Anatomy of a Science Circus, this volume, chapter 16).

There are many different forms of science show.

Science Shows without a Theme.   Quite often science shows have no particular theme and simply offer a series of demonstrations.   Such shows are often of the 'gee-whiz' variety, not linked to everyday situations, and the underlying science may not be explained.   Some even go as far as to call their shows 'Magic'.   They often last for up to an hour.   The University of Utah stages an annual *Michael Faraday Show*, with presenters in costumes of the period.   The title is a cover for a chemistry show.

Science Shows on a Particular Theme.   These shows are of short duration, usually about twenty minutes, and are presented by travelling exhibitions like the Questacon Science Circus as well as science centres.   They are structured around one particular concept–the centre of mass, perhaps.   Such a show contains a number of demonstrations that all relate to centre of mass, linked by a story line.   The presentations may illustrate how the concept is relevant to familiar situations.   They might deal with ideas like the way pigeons walk and why pregnant women have to adjust their gait, how projectiles move and how men and women are topologically different. Other themes might be: propagation of sound waves; liquids, solids and gases; effects of very low temperatures; and the behaviour of non-Newtonian fluids.

They are usually organised so that members of the audience are able to try their hands at some of the demonstrations. The most usual way of developing them is to choose a topic and then collect as many relevant demonstrations as possible. A story line is then created and the demonstrations are slotted into place.

Television Science Shows. A number of presenters have used the television to communicate science. Jearl Walker - of *Scientific American* fame - made a series of whimsical videos that addressed various physical concepts. He treated his audience to a range of moving images that illustrated his talk and addressed topics like the physics of the fair ground and chemistry of the soufflé.

Theatrical Presentations. In recent years a number of longer presentations have been produced, centred around particular scientific personalities. Typical of these are Questacon's presentations on Archimedes, Galileo, and Alan Turing. They address the science but they place it in the context of history, with biographical details of the person concerned. The presenter attempts to make the scientist come alive for the audience. In Australia, Donna Cohen and Gary Fry have staged a number of one person theatrical presentations about famous, and sometimes not so famous scientists. Donna Cohen does a one-hour presentation in which she plays the English fossil hunter Mary Anning. Gary Fry has written and presented one man, one hour shows about Charles Darwin and Douglas Mawson, the Antarctic explorer. In Britain Johnny Ball is well known for his character presentations of famous scientists. Bertholdt Brecht wrote a play *The Life and Times of Galileo*, Heinar Kipphardt wrote *In the Matter of J.Robert Oppenheimer*, Gabriel Emanuel made a theatrical study of *Einstein* and Hugh Whitmore's play *Breaking the Code* covered the turbulent life of the English mathematician Alan Turing.

Science Plays for Children. Over a number of years there was a very successful series of short plays written and delivered to children in London's *Molecule Theatre*. Each play revolved around a particular scientific topic coupled with a scenario aimed to capture the children's attention and imagination. One example, concerned with various aspects of mechanics like levers and friction, was about a robbery during which two thieves attempted to steal a heavy safe. They enlisted the help of the audience in much the same way as in English pantomime. The actors explained what they intended to do and how they intended to do it, and then asked the audience to comment *en masse* for their opinion.

Puppetry. This medium is particularly successful with younger audiences. The technique has been used in different parts of the world and one example is a program that was developed as the lead up to the Edinburgh Science Festival. A team of young puppeteers tours around schools giving puppet shows that are all linked to scientific topics.

8. THE BUTTERFLY EMERGES: QUESTACON - THE NATIONAL SCIENCE AND TECHNOLOGY CENTRE

In the early days, Questacon steadily grew and prospered under the Australian National University's wing. Then the Australian Federal Government took a momentous decision. In 1982 the Australian Bicentennial Authority was formed. The body was charged with the task of planning the 1988 celebrations of 200 years of white settlement.

The Authority decided that an interactive science centre would be completed in time for the 1988 celebrations. Thereafter things happened very rapidly. By 1986 the Federal government had agreed to the proposal and the Japanese nation had offered to make Australia a birthday present of ten million dollars towards the capital cost of the building. The construction of the new building began in 1986 and it was completed on schedule and within budget in September 1988. Questacon – the National Science and Technology Centre - was formally opened by the Prime Minister Bob Hawke on the 23 November 1988.

9. YES, BUT WHAT DO THEY LEARN?

A question I often hear is, 'Yes, interactive science centres are all very exciting but what do people actually learn in a science centre?' Let me state here that it is wrong to infer that interactive science centres are engaged in teaching scientific facts to their visitors. They are not, and were never intended to be formal science teaching institutions. Frank Oppenheimer, once said. 'Nobody flunks a science centre!' Just as art galleries do not teach you how to paint, science centres do not teach you science. The art gallery and the science centre each provide a cultural experience.

One unfortunate side effect of the emergence of interactive science centres was the response that they drew at the time from some conventional museum professionals. They saw interactive science centres as a threat to museums because some people stated that they were better than museums - in museums visitors could only look, whereas in science centres they were actually encouraged to touch things. This comment misses the important differences between the two types of institution and the separate needs that they address.

Museum professionals, commenting from their background of research based institutions, complained that visitors to interactive science centres rushed around pulling levers and turning cranks, paying scant attention to what was happening - they simply messed about. The implied question is what do people really learn in such places? The implied answer is - not much!

In response, I fall back on a line from Kenneth Graham's immortal book *The Wind in the Willows*. Ratty remarks at one point that,

> There is *nothing* - absolutely nothing - half so much worth doing as simply messing about in boats. (p.24, Folio Society Edition, 1995)

I believe that there is absolutely nothing like messing about with science and I know a lot of scientists who agree with me. Until recently, however, very little research into what people take away from science centres had been carried out. Now we know that people are indeed changed by their visits and that informal learning does, in fact, take place.

I am convinced that interactive science centres must be both relevant and entertaining. Unhappily there are those who believe that, by making the science entertaining, it is somehow devalued. It is a strange view: no one would object to the notion of being entertained by an opera, a play or a concert, but science is somehow seen by many as different from other aspects of our culture. The last word in this matter should go to a Nobel Laureate. Konrad Bloch, the biochemist, said in his acceptance speech that 'science is the ultimate in entertainment'.

10. THE BIRTH OF THE SCIENCE CIRCUS

From the beginning, we strove to make Questacon a truly national science centre, accessible to all Australians Even before it left the shelter of ANU, I had carried out experiments which eventually led to the establishment of a travelling science circus. This marked the beginning of Questacon's outreach operations, and it began in a rather unexpected way, but one that is relevant to other developing science centres.

As the number of visitors to Questacon soared, short science demonstration shows were introduced (see above). Many of the show givers were young ANU undergraduates and they took to the activity with great enthusiasm. Shows were a great success and the presenters became very popular with the visitors. However, an unexpected bonus for them was how much and how fast the students developed their communication skills. By 1985 there were more than a dozen ANU undergraduates engaged in presenting science shows to the public and in that year I decided to establish a travelling version of Questacon.

The Questacon Science Circus took to the road for the first time in 1985, in an old furniture van. It began by making short weekend visits to towns within a 100km of Canberra. Initially only about 20 interactive exhibits were toured but in addition there were props for some 15 science shows.

In 1988 the travelling show had attracted major financial support from Shell Australia and the Shell Questacon Science Circus was born. This success led the ANU, with some prompting, to establish a postgraduate course in science communication.

The Shell Questacon Science Circus heralded several other travelling initiatives within Australasia. Currently Brisbane, Adelaide, Wollongong and Perth science centres are all sending travelling programs for both adults and children into the more remote regions of their respective states. Across the Tasman Sea there is the very successful Science New Zealand Road Show.

Another recent outreach project, Science on the Move, was built and operated by Questacon with funding from UNESCO, and the Australian Federal Department of Foreign Affairs and Trade. The exhibition contained 33 exhibits and a number of science demonstration shows, similar to those pioneered by the Shell Questacon Science Circus. In a seven-month tour it was seen by 61,000 people and over three hundred teachers participated in special workshops which were conducted as part of the tour. Science on the Move visited the Cook Islands, Fiji, Kiribati, Marshall Islands, Solomon Islands, Tonga, Tuvalu, Vanuatu and Western Samoa. One student in the Solomon Islands commented that the exhibition beat watching soccer on TV.

Another outreach project from Questacon is the Questacon Maths Centre. This facility originated in Canberra and was successfully operated by the ACT Schools Authority for twenty years before it was taken over by Questacon and transformed into a national facility. So successful has the Maths Centre been that a copy has been set up in one of the Melbourne suburbs. The Questacon starlabs are portable inflatable planetaria, and travel to many remote regions of Australia. Local school teachers are trained in the use of the equipment, so that they can give astronomy courses to their own students.

One of the very important roles for a science centre is promoting public lectures. Sometimes these are staged in parallel with a particular exhibition and provide visitors with a greater insight into the science that is on display. On other occasions the lecture is a 'one-off' affair, and may well take advantage of the presence of an eminent scientist who happens to be visiting Canberra. Questacon has frequently sponsored overseas speakers and has then played its national role by enabling these people to speak at several venues around Australia.

11. LINKS WITH THE FORMAL EDUCATION SYSTEM

Close ties have been forged all over the world between interactive science centres and the formal education system. These links have produced some very interesting and valuable results.

Soon after Questacon opened it drew an unexpected reaction from teachers visiting the Centre. They observed that the interactives were simple and could easily be duplicated and introduced into their classrooms.

It was not long before Questacon was running in-service programs to provide teachers with ideas that would readily translate into the classroom. Chalk and talk has long been the main technique of teaching science, and this has led to much disenchantment among students over the years. In the early 1980's, changes began to appear in the Australian school system and there were many teachers who were seeking new and better ways of teaching science to their students. The opportunities offered by Questacon were quickly seized

This sort of link with the school teachers has also happened in America and Great Britain. The Exploratorium has, for a number of years, run very successful programs for both primary and secondary teachers in the San Francisco Bay area. When Great Britain decided to introduce science into primary school curriculum, science centres like the Exploratory in Bristol and Techniquest in Cardiff were inundated by teachers seeking ideas and help.

11.1. Links with Universities

The links between science centres and universities have already been mentioned in Section 2. To this list can be added the link between the Science Museum and Imperial College in London and the strong links that exist between Glasgow University and that city's new interactive science centre opened in 2001. The University of Mexico has a very similar arrangement; the National Science Museum is part of the University and students who work in the museum as explainers are also involved in a science communication course in the university.

12. ATTRACTING SPONSORSHIP

A major challenge for Questacon right from the start was funding. The Centre attracted, on average, one million dollars a year in sponsorship for the first ten years of existence. Few, if any, of the other cultural institutions in Australia can boast this achievement. The Federal Government requires it, however, to generate half its annual budget. The other half comes in appropriation from the Federal Government. It is the only national cultural institution in Australia that is given this task and it costs around ten million dollars a year to run. Money is generated by charging entrance fees, hiring out all or parts of the building to outside bodies, running a gift shop, charging for professional services and leasing exhibitions to other institutions.

Most science centres have a membership program. The opinions about these schemes depend on which science centre you speak to. Some report that they are more trouble than they are worth. and others say that they get great support from such bodies. My view is that membership programs should not be seen as revenue raising bodies. The most important aspect of members or 'friends', as they are sometimes called, is that they can become extremely effective advocates for science in the community. A wide base of public support in the community is very valuable. When crusading politicians sense such support they equate it with votes and tend to avoid trying to make sweeping changes.

Members of Questacon are mostly family groups who tend to come fairly regularly. One of the attractions of most membership programs is that they offer free admission for a 12 month period. Another attraction that membership programs bestow is reciprocal membership to other science centres and museums. Questacon runs a number of special events for its members. When a new exhibition opens there are special preview evenings for members and they are offered special discounts on merchandise in the Centre's shop.

13. TWENTY YEARS ON!

Questacon has achieved much since its inception in 1980. It has been transformed from a small operation in an unused school into a large custom built science centre in the middle of the Australian capital city, Canberra. It set out, from the beginning, to be a truly national science centre and it has achieved this goal in a number of ways. The average number of people that Questacon reaches each year, in Canberra and around Australia and overseas, is over one million. This is an impressive percentage of a population of only 20 million. There is no other science centre in the world that covers such a vast territory. Questacon has built interactive exhibitions which have toured all over Australia. A lament that is often heard from visitors to Questacon is 'Oh how I wish something like this had been around when I was at school'. Well today, it is!

National Centre for the Public Awareness of Science, Australian National University, Canberra ACT 0200, Australia

16. THE ANATOMY OF A SCIENCE CIRCUS

The Evolution of a Graduate Program in Science Communication

1. INTRODUCTION

> So at the beginning of this year (1998), CV in hand, I knocked on a few promising doors looking for a job–or at least some experience. Doing this I found employers quite fascinated by the qualification I held. What is this course? What did you do? Who lectured you? I was quite happy to invoke their thoughts of my boundless experience from such a strange, mythical year.

This quotation, taken from a letter from Ashley Turner, a member of the 1997 Shell Questacon Science Circus, describes his experience on embarking on a career in technology transfer. The 'mythical year' in question was 1997, when he enrolled for the Graduate Diploma in Scientific Communication. He spent half of his time in Canberra, at the Australian National University and Questacon–the National Science and Technology Centre, and the other half touring Australia with the Shell Questacon Science Circus...

2. THE CIRCUS COMES TO ECHUCA

The vast, blue-grey plain extends to the horizon in all directions. The colour is due to saltbush, a salt resistant and drought hardy shrub that, with spinifex, forms the dominant vegetation in the area. The air shimmers in the 35° heat and parts of the horizon are lost in mirror-like mirages that conjure up thoughts of cool waters. A few dust covered and disconsolate sheep, living on the saltbush, ruminate close to a corrugated iron windmill that dips into the artesian basin and provides a constant trickle of brackish water. Not far away, also attracted by the bore water, a small mob of kangaroos looks relaxed as they lie on their sides in the heat of the day, scratching their stomachs and licking their forearms to take advantage of the cooling effect of any slight breeze as it evaporates their saliva.

This is one of the flattest places on earth. The nearest 'hill' is seven metres high and more than eighty kilometres away. The road heads for the horizon in a line so straight that it might have been laid down by a Roman engineer. Looking down the road is like viewing a painting by a master of perspective. Its edges are not parallel - they meet at the vanishing point on the horizon, accompanied by the telegraph lines that link the distant towns. A painting by Russell Drysdale.

S.M. Stocklmayer et al. (eds.), Science Communication in Theory and Practice, 237–255.

Along this road, at a steady 100km/h, a large articulated truck - a pantechnicon - is eating up the distance as it makes its way to the horizon. It is a rather lurid truck, one side depicting an orange and red day with the symbols of modern science and technology, and a black and blue night on the other with images of astronomy. The driver gets a lot of friendly ribbing from his fellows, by radio or at the road-stops that cater to truckies.

After two or three hours the dry country gives way to eucalyptus woodland and a particularly dense line of river redgums marks the banks of the great Murray River. Houses become more and more frequent, and the pantechnicon slows to the required 60 km/h as it approaches a bridge. To the right, on the river, a small fleet of paddle steamers - once, this river made it all the way to the sea and this was the biggest inland port in Australia - are moored at the wharf waiting for trippers. With a rattle, the truck is over the bridge, turns to the right and after a few minutes pulls into a showground. The Circus has arrived in Echuca.

2.1. Setting Up

The showground is a large open space with a fenced arena and a grandstand. There is a large, utilitarian building - four walls, a high roof and unclad steel pillars. It is a building more used to agricultural shows, craft fairs and community events but today, and for the next week, it is the venue for the ANU Shell Questacon Science Circus.

The students who run the Circus have arrived the previous day. There are ten of them, ages ranging from 22 to 30, and a coordinator. There are more young women than young men. This is a rule which has been established empirically over twelve years of operation, even though the selection process is even handed. It reflects the fact that women scientists tend to be more interested in communication. The coordinator, one of three, is an old Circus hand, but no older in years than the students, if older in experience.

The whole Circus team meets the pantechnicon and at once there is a flurry of activity. The task is to unload some 50 interactive exhibits and set them up in the large hall. They have done this many times before and set about it in the confident, organised way of a well drilled team. There are no specifically assigned jobs - each member of the team knows what has to be done and each puts her or his effort where the immediate need exists. Small groups form to achieve some end, dissolve and reform in different combinations. The felt blankets that act as packing for the exhibits in transit are removed, carefully folded and stowed away. The wheeled dollies are brought out and each blue box containing one exhibit is quickly hefted on and run into the hall. The coordinator and one of the students, the designated floor-manager-for-a-day, decide on the overall layout - where the speedball exhibit will cause least disruption as visitors test the power of their throwing arms; where the harmonograph can be set up with the least danger of being bumped by enthusiastic children; where the little theatre for science shows can be arranged, as far from the general hubbub as possible.

Once the exhibits are brought in, the pantechnicon is parked, ideally in a spot that will signal unequivocally to potential visitors that *the circus is here*. This releases the driver from driving duties and, though he doesn't have to, he plunges enthusiastically into the fray. His contribution is very much appreciated because, while there is an abundance of brains, brawn may be in short supply.

Each exhibit is unpacked from its blue box, which now forms a stand on which the exhibit will be mounted, given a dust and a rub down with a mixture of water, methylated spirit and detergent. The loud speaker system is set up and chairs are arranged in the space chosen for the presentation of science shows. Tables are set up for the various props and safety devices. The travelling shop is assembled. It is a large perspex show-case containing a myriad cheap scientific toys, such as bouncing putty, gyroscopes and gliders. They are very popular and their popularity is enhanced because the students use many of them in their science shows. A table with a cash register is placed by the door where the entrance fees will be collected. Posters containing information about the two institutions that send out the Circus are placed strategically. On a good day all of this takes about 90 minutes.

There is now time for a rest before the venue opens to the public at 6pm. Students return to the lodgings that have been booked for them some months earlier, to sleep or read. The more energetic might go swimming or window shopping or simply sightseeing around the old town which has seen so much of Australia's history.

The students have not arrived cold in Echuca. Some months ago, when the tour was planned one of the coordinators visited the area to make arrangements; to seek out cheap but pleasant accommodation, to check out the schools where teams of two students would perform their science shows and busks for the children, to make sure that there was advance publicity and that there will be good media coverage when the Circus is in town. Planning is a big job as six tours must be organised, at least one of which will probably take place on the far side of a continent as large as North America. The tasks of the coordinators are thus threefold. They must organise the Circus, train the students, and provide pastoral care. All of this is done with the help of the combined resources of Questacon and the National Centre for the Public Awareness of Science at the Australian National University.

2.2. The Doors Open

Thirty minutes before the Circus opens at 6pm, the students assemble for a pre-Circus briefing with several volunteer explainers from the senior forms of the local secondary schools. The Floor Manager ensures that all the science shows are ready to go at their scheduled times and that each student knows when she or he is on stage. The doors open and in comes the public, a trickle at first but later a flood that surges onto the floor. At once the hall is alive with noise - the noise of children dashing from exhibit to exhibit, waving musical pipes and bouncing superballs while their self-conscious parents follow sedately in their wake. The students are now hard at work explaining to the little clusters of visitors exactly what scientific principle is illustrated by the exhibit before them. Two other students are doing a roaring trade in rocket balloons and slime in the shop.

After an hour, the Floor Manager calls for a little silence and announces that the first science show will shortly begin. There is a rush for the theatre area and the floor is left strangely hushed until the influx of new visitors restores the noisy *status quo*. The student putting on the show explains who she is and where she is from and begins. It is the *Liquid Nitrogen Show - the Coolest Show on Earth*. It is a sure winner, producing plenty of dense fog, frozen bubbles and rock hard bananas. Other shows scheduled for tonight are *Balancing the Improbable*, about the centre of gravity, *Up Up and Away*, about the science of flight, a chemistry show called *Fire and Light*, and *The Slime Show*, about non-Newtonian fluids.

At 10pm the Floor Manager starts to encourage people to leave and the door closes on the last of the visitors at 10.15pm. Then there are the takings to be counted and reconciled with the print-out on the cash register, the theatre to be tidied and the floor restored to its original layout. The ritual lottery is held, in which everyone makes a guess at the number of visitors through the 'turnstiles'. The results are scrupulously recorded in a running tally because, at the end of the visit to Echuca, the Circus member whose guesses are closest to the actual tally will receive a small prize. Everyone is then off home to a well-earned supper.

2.3. School Visits

The next day is bright and sunny, though rather cold. Although the Circus does not open until the evening only the lucky few who are rostered off can look forward to a leisurely lie in. The bulk of the students are up, checking their props for their shows and their busks - science minishows that are useful in occupying the minute or two between dismantling one show and setting up the next - consulting maps and warming up the cars. For, during the day, the students take a small part of the Circus out into countryside around the city, to the small villages and hamlets that have their own primary schools. Or they visit the secondary schools within the town itself.

The students set off in pairs, each pair headed for a particular school. On arrival at the school there is an initial uncertainty as they try to find the contact teacher and then again as they try to find the car park closest to where they are to give their shows. The shows are usually given in the gymnasium, the art room or the school hall. They set up their props and wait anxiously as class after class files in. One member of the team entertains the children with a science busk as the other puts the final touches to the set up for her show. The show is a great success, for one of the skills that the scholars have developed is the ability to trim the show according to the age and level of understanding of the audience.

They may be giving two performances at the school. If so, that is a hard morning's work because the children are very demanding even when moderated by their teachers. There is only a short break between performances and that is usually spent in the staff room where the importunings of the teachers are scarcely less than those of the children. Then it is back to the hall for the second performance and, as it is close to lunch time, dealing with all the children who stay on and want to play with the props. Eventually they extricate themselves, load the car and escape to a late lunch.

Even then there may be no leisure, as they have to talk with the media, give interviews (the Science Circus received excellent media coverage on this tour, with a total of over 50 stories in local newspapers, on radio stations and on regional TV stations) or make a special visit to the hospital. At 6pm, it is back to the Circus venue for another four hours of bringing science to the people.

They repeat this exhausting schedule, with variations almost every day whilst on tour. Supervisors of mature years wilt after about two days. It is a job for the young and fit!

3. THE CIRCUS IN 2000

In 2000, the Circus reached a quarter of a million people. It was an unusual year, however, because teams of Circus members accompanied the Olympic Torch on its travels across Australia, entertaining the crowds en route. In a more typical year, the Science Circus plays to 100,000 people, young and old, at venues and in schools, in the course of six tours.

Australia is remarkable in that the bulk of its population is concentrated in a few very large cities; once the cities are left behind, settlement is sparse. A typical annual program is this one, undertaken in 1997. On leaving Canberra, which is set high in the tablelands beyond the coastal ranges, and is the biggest inland city with a population of 300,000, the Circus headed south west. A two or three hundred km drive brought it into the southern tablelands of NSW. Here the climate is mediterranean and most of the towns are supported by sheep and wheat. Its next tour was down to the far south east, about 1000km, to the Gippsland region of Victoria; the weather is colder, but once again the Gippsland communities are sustained by agriculture, sheep and cattle. In complete contrast, the third tour was more than 2000km north, into the tropical and subtropical regions of Queensland and the Torres Straits, into high humidity and everlasting summer. Here, sugar cane, mangoes and other tropical fruit and cattle are the staples. The fourth tour was out to western NSW, 1000km away from the lush coastal farming country and into the saltbush, where it might freeze at night and reach 35° in the hottest part of the day and the air is dry. Rainfall is so low here that farmers commonly run one sheep to two hectares; farming is a very fragile living indeed. The fifth trip, another 1000km, took the Circus to south-western Victoria, to the fruit growing regions of the Murray Valley and to Echuca. Finally, the Circus made its way 5000km across the breadth of Australia, to the Pilbara region of WA, a desolate land of red cliffs and red soil and iron ore.

The students were on tour for a period of 23 weeks but not continuously, as only ten of the fifteen are out at any one time. Each student goes on four tours. A lottery determines which tour they go on and there is obviously great competition to go on the most exciting and distant tour, this time to the Pilbara.

A number of special events occurred whilst on tour. In north Queensland, the Science Circus made special visits to the islands dotted in the straits between Australia and Papua New Guinea. There, it performed for Aboriginal and Torres Strait Islander schoolchildren and communities. In the north of Western Australia, it visited 14 remote Aboriginal communities - and in the Pilbara, the word 'remote' takes on a meaning quite different from its usual one. While in WA it also appeared as part of the FeNaClNG Festival - a science festival sponsored by Lions GWN and whose name is made up of the chemical symbols for iron, sodium and chlorine plus Natural Gas and is probably a pun on 'finarkling', a fictitious game invented by an Australian comedian - and the truck took part in the festival parade.

In the outback, through the School of Distance Education in Queensland and the School of the Air in WA, the circus members presented on-air lessons to students in remote areas. On-air lessons are particularly demanding, as all instructions have to be verbal because, of course, it is radio, not television. Try describing the Moebius strip without using your hands! The Circus gave performances and workshop sessions for students in Special Schools in Victoria, for children in a paediatric hospital ward and addressed Rotary club meetings in a number of towns. They also ran a workshop for young men in a detention centre in Townsville.

As if this was not enough, the students, as part of their course, undertook work placements. Each student spent at least one week in one of the various departments of Questacon and another week on outside placements at, for example, the Australian Broadcasting Corporation's (ABC) television studios in Melbourne, where the Natural History Unit, and *Quantum*, an award winning science show, are made. Other students visited Channel 10's wildlife show, *Totally Wild*, in Brisbane. Other placements during the year were with the University of New South Wales' Communications Branch in Sydney, the National Dinosaur Museum in Canberra; and the thematic Seaworld in Queensland.

3.1. The Coursework

When they are not travelling or completing industry placements, the students return to Canberra to complete the coursework components of the Diploma. There are three major components: *Print and Electronic Media*; *Science and Society*; and the *Applied Project*.

The *Print* component includes three writing assignments. The National Centre for the Public of Science is fortunate to have an eminent journalist as a Visiting Fellow. Under his tutelage students write a news stories and feature articles for critical discussion during a workshop The best pieces are rigorously edited and many are published in newspapers and magazines.

For *Electronic Media*, students are coached in interview techniques by the Australian Broadcasting Corporation's senior media trainer. They spend the first part of the short workshop practising their presentation skills without a camera in sight. Then they transfer to a modern TV studio for familiarisation with the unusual studio environment before occupying the hot seat, one by one, to be videoed while they explained what they do with the Circus. After detailed analysis, students are asked to prepare a short piece about the Circus, using it and its exhibits as background. The results are a series of very polished and entertaining 'takes'.

Science and Society is concerned with the interaction of scientific discovery with important social and ethical issues It includes a number of very thought provoking sessions on the background to understanding and the constructivist model of communication. Issues such as metaphor and analogy in science, medical ethics, and science and gender generate heated discussions. Students later present their own seminars on the interaction of science and society.

The *Applied Project* gives students the opportunity to work as part of a team to design an interactive science exhibition, one that might be built and incorporated into an existing science centre. The brief for exhibition design places emphasis both on the overall exhibition concept and on the very well produced dossiers illustrating the finished display, which will ultimately become part of the professional portfolio of each student.   This ambitious enterprise is supported by seminars from Questacon staff on exhibition development, on the creation of educational materials, on exhibit design, budgeting, graphic design, layout of text panels and dossiers and exhibition support.

4. WHO ARE THE STUDENTS?

The first and most important point to understand is that the students are not people who failed to make the grade in a more orthodox scientific career. Rather, they are highly intelligent and strongly motivated young men and women, many of whom have looked upon a field that has been mishandled by scientific administrators and have decided to create their own opportunities. Many of them have seen how bright students become cannon fodder in research laboratories, especially those laboratories concerned with molecular biology; how after three or four stressful years at the bench they can look forward to three or four more stressful years of post-doctoral work; how only a small proportion of those post-doctoral fellows find more senior employment; and how most of those do not enjoy permanency, or even what passes for permanency in an age of economic rationalism. Instead, they have looked about and sought to equip themselves with a set of skills not readily encountered in science courses, to give them an edge over their more conservative colleagues. And it pays off.

The competition to join the Circus is tough. Around July, the course for the coming year is advertised all over Australia and recruits come from all States (Table 1). They also come from a wide range of disciplines, although the biological sciences are over represented. This is perhaps because, in Australia, the biological sciences are traditionally the preserve of women; the effect is that women have made up nearly 70% of all Circuses. It may also be due to the fact that disciplines such ecology, molecular biology and genetic engineering (subsumed in the table under the headings Biochemistry and Biology) have enjoyed considerable expansion, which has resulted in the creation of many positions for hopeful PhDs but little career structure beyond.

Table 1. Origin and first degree of Science Circus Students (numbers of students, 1988 - 2000)

State of Origin		First Degree	
Western Australia	7	Biochemistry	40
South Australia	31	Biology	79
Victoria	28	Chemistry	15
Queensland	22	Computer Sci	4
Tasmania	6	Engineering	3
New South Wales	52	Mathematics	10
Aust Capital Territory	28	Physics	21
Northern Territory	1	Geology	3
New Zealand*	3	Psychology	4
United Kingdom**	1		

*Australian resident; ** Full fee paying student

The advertisements call for 'expressions of interest' in one of 15 scholarship places with the Circus. Application packages are sent to those interested; there have been as many as 800 such expressions, which have translated into about 100 applications. As it is a scholarship course, emphasis is put on academic achievement in the first instance, with the effect that many people, who have only mediocre results, are dissuaded from applying. The remaining group positively scintillates with glittering prizes, first class honours and university medals. The pass students who apply have records studded with distinctions and high distinctions.

A short list of about 50 is constructed from this feast of talent. Each one is sent a video of a science show being delivered by a trained presenter. They are also sent a script and the instructions for making the simple props necessary for the show. They are asked to make a video, a science show of their own - they can depart from the script if they wish - and return it for audition. They are enjoined to use an audience, because shows work better with audience participation. They are also required to use a stationary camera. We are interested in the skill of the presenter, not that of the camera operator.

After the appointment committee have viewed the videos - an often hilarious day due to the inventiveness of the applicants - a final short list of fifteen plus three reserves is established and offers are sent out. In the first ten years of operation, only two or three people in the short list have rejected the offer. In the eleventh year, with changes, imposed by the Australian Government, in eligibility for grants for postgraduate study, four turned down, albeit reluctantly, the offer of a place.

This happened in spite of the fact that entry into the course offers a modest scholarship (paid from the interest accruing to a capital sum that was accumulated painfully in the days before 1988, when Questacon operated, in larval form, under the aegis of the ANU in a disused primary school in Canberra) and full expenses, paid by Questacon, whilst the Circus is on tour.

There is an important lesson to be learned from this. Even programs, such as this one, supported by influential people, from Nobel laureates, science ministers and the vice chancellors, downwards, can suffer 'collateral damage' as a result of policies directed towards quite different targets.

4.1 What Happens to the Circus Veterans?

We keep in touch with graduates by means of the in-house newsletter *Scinapse*. *Scinapse* was started by the students themselves a decade ago, and although it is now edited by the present author, Circus veterans are the major contributors. It contains news of past students, of the Graduate Program and the National Centre for the Public Awareness of Science, provides information about jobs and publishes comment about science communication as well as the occasional serious, short article. As a consequence, information about the careers of the graduates is as up-to-date as that for any other student cohort–and probably better than most.

Table 2 shows the first employment taken up by students on completion of the Diploma. Even though we claim to have good information, it is still difficult to keep track of such a volatile cohort of students so the following table gives the percentages of graduates employed in various categories of work. Inevitably we lose track of some, and these appear as 'unknown'. The number of 'unknowns' is also swelled by the previous year's graduates, many of whom take a break before seeking employment.

*Table 2. First employment of Science Circus Students
on completion of Diploma. Figures are percentages of
total enrolments 1988 - 2000*

Sci. Comm. 'Industry'	59
Industrial Research	3
University Research	1
Higher Degrees	12
Secondary Teaching	6
Other	2
Unknown	17

Only the category 'Science Communication Industry' really needs explanation. This includes journalists, TV and radio presenters, workers in science centres and museums, communication officers for scientific, environmental and industrial establishments, professional associations and exhibition designers. As this is by far the largest category of employment, we can claim satisfaction from the thought that at least we do not appear to have done the students any harm! Follow-up suggests that most continue in the field and some from the other categories move into it.

Of the other categories, it is clear that relatively few students resume research careers on completion of the course. This is hardly surprising as, first, it was not meant to be a preparation for research but to add value to the existing qualifications of the students and second, many of the graduates join the Circus because they are disenchanted with research and its prospects. A significant proportion (12%), however, do resume an orthodox graduate career in science, enriched, we hope, by the experiences of a very unusual year.

4.2.. What Do the Scholars Think?

It is one of the great strengths of the program that it has so many outspoken graduates. In 2000, Questacon and the National Centre for the Public Awareness of Science looked into the future in order to design a Circus and a Graduate Program, of which the Graduate Diploma is a part, for the new millennium. We have turned to the Circus veterans themselves, to inform and instruct us about the Circus in 2001 and beyond. We maintain a Circus emailing list, which includes more than 90% of veterans, and have started a debate about the future. As might be expected, there are both positive and negative views in the responses of the veterans and many suggestions for change. Here I summarise some of the points that have been made about the Circus and the Graduate Diploma Program in the past. There is, of course, much advice about the future as well but that must wait for another book!

Touring has for most been one of the great strengths of the course but, obviously, there is too much for some. In fact, over the ten years of operation, we

Touring has for most been one of the great strengths of the course but, obviously, there is too much for some. In fact, over the ten years of operation, we have increased the number of students from nine to fifteen and reduced the number of tours undertaken by each student from six to four. Touring is seen as both a strength and a highlight that

> creates a clientele with a long memory...provides a source of diverse and clever communicators for elsewhere. (1989 veteran)

This is from an early Circus veteran who now employs people to run a science section in a museum. But his view is not universal:

> Firstly, I also think that there was *way* too much emphasis on touring, though I gather current scholars do much less than the one mini (tour) and five major tours that we did. I truly believe that, after the first few weeks of explaining and giving shows, the returns obtained significantly diminish *unless* you have a passion for performance. (1990 veteran)

while a 1992 veteran points out that

> The most obvious conflict is between touring and more conventional coursework. Both are extremely valuable for different reasons and I don't think either should be compromised any more than they already are.

The Circus also provides an important service to rural communities, whose remoteness leads them to value personal contact.

> ...being in a rural location I can tell you that teachers and students (and the greater community) certainly do enjoy the personal contact a travelling program can provide. While overlaps between the various programs do seem to exist, there is a real need for touring programs in the really remote areas of the country. (1996 veteran)

There is common feeling for the uniqueness of the course and its breadth - although a few lamented the sacrifice of depth - that enabled one scholar to 'develop as a person not just a communicator'.

It was seen to be a great avenue for networking within the science communication group and for promoting understanding that science and technology not only concerns the advances and achievements in science but how science interacts or fails to interact with society.

Of course, the success of the program is due to the efforts of the students themselves and their subsequent achievements..

> The program is too valuable an enterprise to lose. The experiences I have gained from the year have helped me enormously. I would not be doing what I am today if it wasn't for the circus. (1996 veteran)

Many comment on the broad range of experience provided by the Circus and see it as one the great strengths of the Graduate Diploma Program

> I find I have used the concepts, skills and experiences gained during the Grad Dip throughout the various jobs I have had, both in science communication and in public health, and I will continue to do so. (1992 veteran)

> The one thing I learnt from the circus, is the many things that were possible to learn in the circus. Just when I think I have used all my sci comm skills, another challenge arises and I find myself able to reflect on a 1998 experience to help me through. (1998 veteran)

Not all comments are favourable, however.

> However I think that in trying to cover everything and please everyone, the Grad Dip stretches itself too thin. (1998 veteran - different from the above)

Team building was seen as an important aspect of the program.

> As for team spirit, touring brings out the best and worst of everyone. It's a must for building strong relationships. .....There are many decade long friendships from the Circus....Heck, we have marriages! (1989 veteran)

Access to the national and international science communication networks is highly valued. The good reputation created by graduates who travel overseas and work for short periods in other countries is something many want to partake in and to which they want to contribute.

> I think that one great strength of Questacon and the Circus is that it has built up a good reputation around Australia and overseas, many audiences will return to the circus based on positive experiences in the past, many ex-scholars will be employed based on their experiences in such a unique post-graduate course. (1992 veteran)

The team work, the interpersonal bonding and the sense of doing a worthwhile job at an important point in the process of acquiring maturity creates a memorable year for the students.

> I would agree with everyone else that a year with the Grad. Dip/ Circus is one of the most memorable events in my life so far and I know I would not be where I am now if it wasn't for that year. (1998 veteran)

> I think anyone who has been lucky enough to spend a year in the circus would wave the flag and proclaim the value of the program for scholars, for the country, for Science. (1994 veteran)

> I look back at 1997 with immense fondness (quite possibly far too much; I recognise the danger that I will someday become an overly nostalgic bore), and most of that stems from the touring. (1997 veteran)

4.3. What Does the World Think?

The year 2001 is a critical one in the life of Science Circus. Shell Australia completes the third five year cycle of sponsorship. If sponsorship were to continue then an external, objective assessment of the program was necessary. Two independent reviews were carried out.

The first review is entitled *Report on the Influence on Graduates' Careers of the Graduate Diploma in Scientific Communication at the Australian National University* (Lucas, 2000)and is available on request. Its salient findings are as follows;

- The Graduate Diploma in Scientific Communication course is highly regarded both nationally and internationally.
- It has an excellent record in terms of graduates securing employment in science communication or related fields.
- Graduates' careers are well served in the short to medium term by the present course structure and content, whether those careers be in science centres, journalism or more generally in the field of communication.
- There is a high level of acceptance by graduates of the range of activities that comprise the current course.
- Graduates of the course are engaged in a striking range of careers in Australia and overseas. A significant minority of graduates are not currently employed in science communication, although, of these, many acknowledge the positive contribution of the course to their current employment.

The second report was entitled *Evaluation of the Educational Effectiveness of the Shell Questacon Science Circus Program* (Rennie and Williams, 2000) It noted that the SQSC members were a friendly, united team with great cameraderie, and identified this an important training aspect for Australia's young scientists. It went on to say

> The results...indicate that overall, response to the SQSC is very positive. Visitors and teachers comment on the professionalism and enthusiasm of the presenters; the portrayal of science as fun, interesting, relevant to everyday life, done by normal people, and accessible to students; the enjoyment of visits by students and the public, particularly the hands-on aspect; and the opportunity for students and the public in rural regions to see and do things they might not otherwise have experienced...the following recommendations seek not to alter the direction of the program but to ensure *that it stays at the cutting edge of international science education.*' (Italics those of the author of this chapter).

5. THE EVOLUTION OF THE SCIENCE CIRCUS

The Science Circus owes its existence to the vision of the Founding Director of Questacon, Dr Michael Gore. Out of a simple idea, to establish a laboratory in which the general public, especially children, could undertake a voyage of exploration and personal enlightenment, has grown both Questacon and the ANU's National Centre for the Public Awareness of Science (CPAS). The story of the Questacon is told in chapter 15, but the events of the last decade that have resulted in the CPAS today–a vigorous if small academic community of more than 40 graduate students (including the 15 Circus students), four academic staff and administrative support, a workable budget and the approval of senior officers in the ANU, are worth recounting as its history is inextricably linked with the Circus.

It would be nice to say that this has been the result of foresight and careful planning. It is not so, I am afraid. It is noteworthy that, at conferences concerned with science communication and adult education many delegates ask questions about the pedagogical objectives of our Graduate Program. When we began to think about a graduate certificate in 1987, this is not where we intended to be in 2001.

science communication. If we did so, it would, unfortunately, be a *post hoc* rationalisation. The genesis of the program lies in Michael Gore's good idea for an outreach arm of Questacon, that is to say, a science circus. The goals of the program are simply stated: to produce knowledgeable and effective science communicators.

Given the calibre of the graduates with whom we are privileged to work, we have been able to adopt a collegial style of operation, dimly recollected from student days in the 1950s, when tertiary education was a privilege hard won and class sizes were a fraction of what they are today. We have allowed students to define their own objectives and have attempted to facilitate their achievement by providing a stimulating academic environment and gentle guidance. If we have a philosophy it is this. We are all equals engaged upon the same enterprise; the difference between us is that some of us are further - much further! - along our career paths than others.

So here we are, owing more to contingency and less to foresight. Had you asked us in 1988 what our objectives were in starting up a graduate certificate, the answer would have been that we were concerned about the viability of an outreach project of Questacon. Had the 1988 students not staged a minor palace revolution and insisted on more theoretical content in the certificate course; had the 1990 Scholars not pointed out vehemently that a certificate was a poor reward for a year's very hard work, catalysing its replacement by the Graduate Diploma after two years; had not the Dean of Science in 1991 *insisted* that we establish a Masters course; had the staffing mix been different, we might be something else.

Our motivation, the belief that science communication is a good thing and that there should be more of it, has never changed. But all throughout the last 15 we have been buffeted by circumstance. As organic evolution is unpredictable, depending on random events, so has it been with the evolution of the Circus. If Dr Gore and the author of the present chapter had not been sitting on a seat on the south coast of NSW, overlooking the sea, drinking beer and discussing science centres, if we had not been concerned about the science circus, if we had not both been in relatively powerful positions so that we were able to further our plans, if the Scholars had not twice galvanised us into change, if my successor as Dean of Science had not been beguiled by the thought of the dollars in equivalent full time student units and full fee paying students and, most important, if we had had these ideas in 1978 rather than in 1988 when we were poised to catch the wave of enthusiasm for science communication in the nineties, if... well, the ifs can be listed indefinitely.

6. WHERE DOES THE SCIENCE CIRCUS FIT?

Elsewhere in this book it is argued that successfully increasing public awareness of science contributes enormously to social well-being as it creates a community that

Elsewhere in this book it is argued that successfully increasing public awareness of science contributes enormously to social well-being as it creates a community that is both confident in its possession of scientific ideas and is comfortable about raising children to have the same confidence.

Now, how does this illuminate what the ANU Shell Questacon Science Circus is attempting to do? Fifteen years ago the philosophy of the Circus reflected the philosophy of Questacon itself which was summarised by what is reputedly a Chinese proverb 'when I hear, I forget, when I see, I remember, when I do, I understand'. But is this what actually happens?

At the beginning of the 21st Century, there is burgeoning interest in the affective changes that informal learning environments bring about in the community. In particular, a study by Gilbert and Stocklmayer (1999) demonstrates that each visitor takes away something different from such places. It is becoming apparent that the task of science centres is not necessarily to 'educate' the public. That can be done but it requires a very specific sort of organisation to achieve it - for example, linking displays to curriculum. Neither is there much evidence that the Science Circus contributes to a significant increase in the systematised scientific knowledge of its visitors. What it does is to contribute to an affective, attitudinal change in its clients. If the Circus does its job properly that change will be positive with respect to science. But there are many people who take away something never intended by designers - for example, Gilbert and Stocklmayer (1999) describe a visitor who has returned many times to the Questacon 'black hole' exhibit. It is simply a funnel with an increasing gradient towards a hole in the middle. Any ball launched eventually makes its way down the hole. But this visitor is not concerned with black holes–what fascinates him is the trajectory of the ball and how that can be changed by varying the delivery. The new knowledge he goes away with is far removed from that intended, and he obviously enjoys himself.

This phenomenon is particularly true if the visitor is a child. It seems unlikely that a five year old visiting the Circus goes away with any clear idea of the liquefaction of nitrogen at low temperatures. But the child will see interesting tricks done with liquid nitrogen, and satisfying fog, and frozen bubbles, and go away with a positive experience and - and this is a pious hope in the absence of detailed information - a heightened sense of wonder.

I have called these little experiments 'tricks' despite what purists would want to say, for the word to them carries associations of magic and conjuring, both of which should be distanced from scientific phenomena. The public, however, thinks they are tricks and, indeed, are encouraged to think in those terms by the very name 'Circus'. A tulip frozen in liquid nitrogen and then shattered like a delicate wineglass is certainly perceived as a trick by many of the audience. There are many definitions of trick in the dictionary, and I use the word in the following sense

....a clever and skilful expedient; a special technique; the art...of doing something cleverly or skilfully. (Shorter Oxford English Dictionary).

The Science Circus attempts to bring about an affective change in its public by a variety of means First, simply by its presence. The Circus venue is brightly coloured, it is clearly an arena where the visitors are having fun. The children are boisterous and parents are pleased that they have an opportunity to play. And it is cheap, which is especially important in outback Australia, which, with the collapse of commodity prices, is in recession. The Circus played to about 100,000 people last year and observation suggests that the vast majority had an enjoyable experience.

Second, by the the fact that it is run by a group of highly personable young people, predominantly women, who are accomplished scientists.

Third, by the shows. Each show is a series of tricks illustrating a central theme. The tricks are meticulously explained especially if they are counter-intuitive. For example, in *The Collision Show*, volunteers are asked to throw a superball on to a table so that it bounces up and hits a board, bounces back to the table and then on. In fact, the ball acquires spin in such a way as to send it back towards where it came in, to the astonishment and amusement of the audience. This is carefully explained and then the ball is lubricated with water and the volunteers succeed in the task.

The show is a show first and foremost and only secondarily is it an educational experience. The aim is to induce a feeling of comfort in the audience when it comes to thinking about simple science. Shows are staged at the venue in rapid succession, but the Scholars also visit local schools in pairs, and perform two shows.

A type of minishow that is very useful for filling in between performances is the busk. Busk is defined by the Shorter Oxford English Dictionary (1993) as

> to play music or otherwise entertain, for money in public places

and busker as

> an itinerant musician or actor, especially one performing in the street.

'Itinerant' captures the Circus mood well, and the team develops busks to entertain children between shows, or to advertise their presence in a town, or to amuse queues waiting for the venue to open. Some are extremely funny; all are extremely ingenious. In the former category is *the waggle dance of the honey bee*, developed by scholar Danielle Quinn in 1999. It involves her dressing up as a bee and then performing the dance to indicate to the audience the direction in which lies a hypothetical source of nectar. In the latter category is *the making of a pizza*, by scholar Jodie Varnai which explains the role of fermentation in making the bread base, the salami and the cheese.

7. ISSUES: SO YOU WANT TO START A SCIENCE CIRCUS?

There is probably no recipe for the creation of a successful science circus. There are a number of things that are necessary but not sufficient because the environment in which a Circus operates will have profound effect on the way that it develops.

The most necessary initial ingredient is a person with vision and energy, someone with a silver tongue who is prepared to do the hard slog of convincing people that science communication is important, that it is valuable and that it is the one thing that they want to invest in. That person must also be lucky. In the case of Questacon, the luck came in the form of the bicentennial celebrations of the settlement of Australia (Chapter 14).

Vision and luck are not enough by themselves. A sponsor is also necessary. A primitive form of the science circus operated from the larval Questacon and was sponsored on an *ad hoc* basis by Shell Australia. When the National Science and Technology Centre opened its doors in 1988, Shell Australia generously supported the Science Circus for the first five years of its existence. The partnership has been so successful that the Circus has just entered on its third five year cycle of sponsorship.

Sponsorship is therefore an essential component of the mix. Other important components proved to be the politics and geography of Australia. The Australian Capital Territory is not part of any State. It is the seat of the Federal Government. Institutions in the ACT often have the prefix 'National'; the National Library, the National Art Gallery, the National University and - the National Science Centre. This prefix implies that the institution is designed to serve the whole of Australia, not just a single State. With the whole of Australia as its territory the Science Circus has been able to range very far indeed. Australia is the size of North America. Many people living in the outback do not get the opportunity to visit a capital city, so they welcome the Circus with open arms when it is in their vicinity. In all the years of its operation it has set up in more than 300 different localities across the length and breadth of the continent, as well as visiting some localities several times. Its brief is to concentrate on rural Australia, on the grounds that urban Australia has access to many more facilities. It has played to more than a million country people. The expansive geography has been a major factor in the success of the Circus. Now, as more States develop their own travelling science shows, it will be interesting to see whether territorial issues become important.

Finally, the most important ingredient has been the students themselves. Because the circus was fortunate enough to be able to offer scholarships it has had its pick of some of the brightest young scientists of each year. Amongst their general characteristic has been the possession of extraverted personalities, widely diverse interests - many of them have been expert musicians, for example - and an intelligent appreciation of the need for teamwork, for professionalism. Managing such thoroughbreds has made pastoral care a challenging and rewarding experience.

8. CONCLUSION - BEST PRACTICE?

For all of the above reasons it is difficult to be categorical about best practice. For much of its existence, the ANU Shell Questacon Science Circus has been unique. Although there are now many travelling science shows, the Circus was, until recently, the only one born of a collaboration between a university and a science centre. Professor John Beetlestone (Founding Director of *Techniquest* in Cardiff, UK) was so impressed with the synergy achieved by the close relationship of the two institutions that he has drawn on it for inspiration in the creation of a program with the University of Glamorgan, and freely acknowledges the debt (personal communication). Another circus is in the making in Queensland, Australia, where Dr Graeme Potter, of Sciencentre, is developing a graduate program jointly with the University of Queensland. In a former incarnation, Dr Potter was associated with the ANU Shell Questacon Science Circus in his role as education manager

A sample of three is too small to allow us to proclaim that the association of a science centre with a university is best practice and there are, of course, many who would deny it vehemently. But if the Circus is associated with graduate training many things become possible. These include the recruitment of young scientists, lacking nerdish qualities, with a passion for communication, to act as role models for young people. There is still surprise among the audience when an attractive young woman stands up to give a show and proclaims that she is a scientist or an engineer.

The rewards for the students transcend the mere financial - which is fortunate! They are trained and at the end of the year receive a qualification which is highly regarded - judging from the readiness with which they find employment - all over the English speaking world.

Letters from Circus veterans are often published in *Scinapse*. Here are two comments about the course (the year in brackets is the year of the writer's involvement in the Circus):

Tracey Bryan (1990) described her life in the Royal Australian Air Force thus

> Apart from School Of The Air, my more general public speaking skills have stood me in good stead. As an officer, I frequently presented briefs and lectures, and certainly had a head start on my peers. My oral and written communication skills have been consistently highlighted as strengths in my annual performance reports.

Andrea Aird (1995) reported on her experiences in England.

> The Circus has really opened doors over here in the UK. It's actually been quite easy finding work. Although, you can't wait for work to fall in your lap, you have got to take the initiative.

Two examples must suffice, lest this page give the appearance of a series of testimonials. There are many more instances. The students themselves understand that the sheer eclecticism of the course, and their experiences on tour, are marketable in a market where they have to stand out as different from the average graduate to have a chance of a satisfying career.

Another great advantage of the association between the ANU and Questacon is that each set of students is with the Circus for one year only. While the students are exhausted at the end of it, they are not jaded nor overcome with ennui. This is not to say that the staff in other travelling shows are so afflicted; simply that each year, the Circus begins with the boundless enthusiasm of young people embarking on a brand new adventure. The Circus is new to them and they show it off to others with the pride and enthusiasm that someone else might evince for a new house or a new car.

All of these things have come together to make the Circus what it is. Not much of it was planned, and if one were to start again, even with foreknowledge, it is doubtful whether it could be reproduced in exactly the same form. Thanks to 15 generations of committed students it has a tradition now, something that cannot be injected at the start, and a momentum of its own. At the beginning it was like tending a baby animal. It was fed and given a good environment and nurtured in a hundred other ways, but in the end it carries the genes for its own growth. A baby elephant will never grow up to be a mouse. We are still waiting to see what the Circus will grow into.

National Centre for the Public Awareness of Science, Australian National University, Canberra ACT 0200, Australia.

REFERENCES

Durant, J.R., Evans, G.A., & Thomas, G.P. (1989). The public understanding of science. *Nature (Lond)*, 340, 11-14.

Gilbert, J.K., & Stocklmayer, S.M. (1999). Mental modeling in science and technology centres: What are visitors really doing?. In S. Stocklmayer and T. Hardy (Eds), *Proceedings of the International Conference on Learning Science in Informal Contexts*. Canberra: Questacon, 16-32

Lucas, K.B. (2000). *Report on the Influence on Graduates' Careers of the Graduate Diploma in Scientific Communication at the Australian National University*. Brisbane: Queensland University of Technology and the Queensland Sciencentre.

Rennie, L.J., and Williams, G.F (2000). *Evaluation of the Educational Effectiveness of the Shell Questacon Science Circus Program*, Perth: Curtin University of Technology.

V. ALTMANN, M. TAMEZ AND D. BARTELS

17. LEARNING BY BUILDING (DESTROYING AND TINKERING, TOO): A POWERFUL SCIENCE COMMUNICATION TOOL

1. INTRODUCTION

It is often said in discussions of education: 'If I hear it, I forget it; if I see it, I remember it; and if I do it, I understand it.' To that we would like to add a fourth maxim: 'If I build it, I own it.' That is, I create its meaning for myself and as a way to communicate it to others.

Building may be one of the most powerful learning methods of all. It is a fundamental feature of many disciplines, including science and art. It is especially compelling because it transcends so many boundaries that seem so intractable in more formal learning settings, such as differences in language, race, gender, and culture. Building is a quintessential human act.

In his book *The Making of the Atomic Bomb*, Pulitzer Prize–winning author Richard Rhodes identifies two types of physicists: theorists and experimentalists. Both were essential to unlocking the secrets of the atom. But it was left to experimentalists like Enrico Fermi and Ernest Lawrence to prove the theorems and conjectures of theorists like Albert Einstein and Niels Bohr.

Besides working in different domains, however, theorists and experimentalists also worked in different environments. The theorists preferred chalkboards, mathematics, and paper; the experimentalists loved workshops, machine tools, and materials that left grease under their fingernails. The experimentalists found supreme joy in overcoming technical difficulties to prove or disprove the popular theories of their day. It is a small wonder that they did not bring down the classroom buildings and athletic stadiums that served as makeshift labs in places like the University of Chicago and Columbia University in New York.

Frank Oppenheimer, founder of the *Exploratorium* and younger brother of Robert Oppenheimer, Director of the Manhattan Project, fell clearly into the experimentalist camp. At the *Exploratorium* - a hands-on museum of science, art, and human perception in San Francisco - the machine shop was then, and is now, the heart of the place. The smell of grease, the screech of lathes, and the sight of sawdust wafting through the rafters assault visitors near the museum's entrance. An old wooden sign hanging nearby says, 'Here is being created the *Exploratorium*, a community museum dedicated to awareness.'

Frank insisted that the scientists, artists, and exhibit builders at the Exploratorium work together in the exhibit-development process. He believed that knowledge would be revealed and created in the design and prototype process itself.

S.M. Stocklmayer et al. (eds.), Science Communication in Theory and Practice, 257–268.

> We recognize that it is essential that neither I nor the staff are bored by our exhibits, that
> we learn something as we make them and that we enjoy showing them to people,
> especially our friends and colleagues, over and over again.

Then as now, the museum's teaching staff believed in the learning benefits of designing exhibits (essentially science experiments) and encouraged teachers attending institute programs to build their own experiments and devices in the museum's workshops. During institute programs, teachers often created miniature versions of existing *Exploratorium* exhibits. In the process of building these 'mini-exhibits', they developed new understandings of different aspects of science, learned new skills, and experienced the discipline, patience, and rigors involved in developing an intimacy with a variety of phenomena.

Today, this notion has extended to children's programs as well, giving students as young as eight a chance to build their own exhibits and work in cluttered, shop-like spaces at the *Exploratorium*. In the process, we have also learned the value of taking things apart and tinkering with them as a means of trying to figure out how things work - even if that means pulling something irrevocably apart, or modifying it to see a change in its behavior. Gadget dissections are a hit for children of all ages (including those in their sixties and seventies!). As it turns out, this approach to learning has a universal appeal.

This chapter is in three sections. The first illustrates examples of this type of experiential learning approach. It gives the reader some idea of the excitement and enthusiasm for science and learning that it can generate. In the second section, we distil some of the lessons we've learned and offer practical advice for those who might dare to attempt this approach with their own students or audiences. We highlight some of the tricks of the trade that help us ensure success and safety. Finally, we offer some ideas about why this approach seems to work so well, and why it is so effective across boundaries of race, gender, age, and culture. Although these are informal hypotheses, we hope some day that researchers will test them empirically, in the best sense of the experimentalists' tradition.

2. EXAMPLES OF THE 'BUILD-IT-YOURSELF' APPROACH IN PRACTICE

Living in San Francisco gives you an insight into why the United Nations originated here: it is one of the most diverse cities in the world, enriched by people from many different cultures, and who speak many different languages. Learning science can be daunting to students who cannot understand English. But experiencing science by doing and building is a great equaliser. Students from other cultures and countries often have more experience in doing things with their hands than their American counterparts. We don't want children just learning vocabulary; we want them to create their own language from their own experience.

2.1. The Mission Science Workshop

The *Mission Science Workshop* in San Francisco is a place where neighborhood kids and sometimes their parents go to tinker and explore. Its founder, Dan Sudran, is a faculty member at City College. He was inspired by the exhibits and the learning culture of the *Exploratorium* and wanted to bring those experiences to the people of San Francisco's Mission District. The facility (altogether, approximately 300m²) is divided into three main sections: an *Exploratorium*-like setting filled with interactive exhibits, a small but useful library and, most important, a workshop.

Professional scientists often begin their careers working in garages and basements, tinkering, taking things apart, and sometimes putting them back together again. *Mission Science Workshop* is a place that facilitates this type of rich and utterly important childhood experience. It is a place where girls and boys can develop a love for science and technology; and gain confidence in their skills. Most of its clients (children) come from poor households, where this type of space and science mentorship are not available. In addition to its work with children, Mission Science Workshop also works with teachers of kindergarten through year 8, connecting informal science to the formal science of the classroom.

Thanks to the success of the Mission Science Workshop, there are now ten additional Community Science Workshops throughout California that were funded by a significant grant from the National Science Foundation. The directors of these sites come from California's diverse population, drawing from different experiences, including backgrounds in construction, museums, and community organisation. Only a few have formal teaching experience.

The Community Science Workshop sites, all of which embody a spirit of tinkering and exploration, have one underlying philosophy, and that is to be child-centered. Most of the activities are child-determined and child-driven. Children who drop in typically spend two or three hours there, two to five days a week, working on individual projects. At first, they do a few skill-building activities. Most newcomers, for example, make small electric motors for their first project. This experience helps them learn how to make a simple circuit, solder, use a hot-glue gun, and measure and cut with a hand saw. Ultimately, the students are exposed to the basic principles of how electric motors work.

After working on a few of these basic activities, the students invent their own projects. At one site, a young boy named Mario had been collecting parts from electronic components and making small inventions from them for years. One day, while we were helping a little girl with a science fair project there, Mario came by and noticed that the girl needed a solenoid. In his four years of taking things apart, Mario had become quite an expert on solenoids and had accumulated several dozen of them. This encounter gave him the opportunity to share his experience - and his collection - with someone who needed his new-found expertise.

At a workshop in San Jose recently, an 11-year-old Latino student who was barely passing his science class used soda cans to create a gearing system for small electric cars. This design is now being used by over a dozen sites in California, and the boy is getting As in science. In another Workshop, one group of children, ages ranging from 8 to 14, had not yet learned to read or tell time. They did, however, design rockets that flew hundreds of feet in the air and built pinhole cameras from old mailing tubes. A few years ago, a seventh-grade girl in the same Workshop fell in love with photography. Now she's doing serious photography in high school. These are but a few of the hundreds of examples being played out all the time at these sites.

2.2. Children's Education Outreach

The *Exploratorium*'s Children's Educational Outreach Program was started in 1985 as a response to requests from public libraries and other neighborhood centers for after-school and Saturday learning for children. After more than a decade, this program has become a major link between the *Exploratorium* and community-based organisations serving inner city children, teens, and their families in San Francisco and Oakland. The Outreach Program offers *Exploratorium* exhibit-based educational activities through ongoing partnerships with these organisations. Program staff go out to the neighborhoods on a regular basis, and also invite participants to come to the museum for special field trips and, in some cases, extended study. All services are offered free of charge.

From the very beginning, the Outreach Program's intent was to give each child an educational experience that included hands-on experiments and activities, as well as the opportunity to build something to take home for further play and exploration. We hoped this might encourage children to keep learning - or, at least, to keep questioning. Creating something tangible that also does something scientifically interesting is very self-affirming for children.

This 'make and take' approach started with simple ideas - creating 'flipsticks and 'flipbooks' during visual-perception workshops, for instance, or making small foam gliders at workshops on aerodynamics. In time, the Outreach Program took its cue from another *Exploratorium* program, the Teacher Institute, which does in-service workshops for middle and high school science and math teachers. Each summer, teachers participating in Teacher Institute programs swarm the machine shop at the *Exploratorium*, creating and building classroom-sized exhibits to take back to their schools. The Outreach Program staff started doing the same with middle school students, giving them the opportunity to learn woodworking, metalworking, and electronics skills while they created their own educational materials.

In 1991, the Outreach Program joined forces with *Aim High*, a free summer school for low-income students affiliated with the San Francisco Unified School District. *Aim High* strives to give middle school children a high school campus–like experience with small classes, afternoon elective courses, and a wide variety of social and cultural programming. As one of the afternoon electives available to *Aim High* students, the *Exploratorium*'s Outreach Program offers 11-, 12-, and 13-year-olds - many of whom come from immigrant families - a five-week intensive institute constructing projects in the *Exploratorium*'s machine shop, working and experimenting together, and exploring related exhibits on the museum floor.

For the children, it's clear that the best part of each summer is building things. Building demystifies science. It helps kids understand the nuts and bolts (sometimes literally) of how things work. A student can build a model car from a kit, but building a model car entirely from scratch is far more satisfying and more conducive to learning. In addition, giving students the responsibility of using real tools and machinery, treating them like adults, and having them work in the machine shop alongside professional teachers and exhibit-builders allows them the opportunity to rise to the occasion, working maturely together, and staying focused and safe. Not only do these learning-by-building situations allow students to take pride in their work, they also serve as catalysts to get kids excited about solving problems and learning science. Primed with positive experiences, we've found that these students are more likely to succeed in school.

For the middle school children we work with, gender equity is often an important issue. Adolescence and early teen years can cause conflicts between boys and girls. We have always had both male and female staff members working with the children, and have been careful to be inclusive at all times. At the beginning of each summer institute, we usually find the boys acting tough and confident about working in the shop, while the girls are more timid. But doing science and math and working with tools are not male-only areas. We train everyone on the tools and machinery in the same way. Once they realize that focus and attention to safety are all that is required to make the work enjoyable and satisfying, the girls quickly become as confident as the boys pretend to be. Occasionally, we see the more confident girls helping out some of the boys who do not yet have the hang of using a cordless drill or belt sander.

Thirteen-year-old Betty, for instance, seemed to learn things effortlessly. She loved building and using tools, but spoke often about the fact that she did not feel comfortable telling her dad about the details of her summer science workshops because, even though he was a contractor, he didn't think girls should be working with tools. Like Betty, Jessica and her best friend Anna, both 12-year-old Asian girls, had experienced a certain amount of cultural bias at home and in their own community (as do most girls) regarding building things in a machine shop and working with power tools. They had convinced themselves that they could not learn to use the tools. One of our female staff members had to coax them to use a band saw for the first time. Both girls shook their heads and seemed scared to try. But by the third week, they were elbowing the boys out of the way to get to the shop machinery. While working on their motor cars, some of the boys quipped, 'Hey! these cars would look cool with working headlights!' At that moment, we saw Jessica and Anna look at each other. The boys didn't follow through with this moment of inspiration, but two days later those girls had working headlights on their cars. Everyone wanted to know how they had done it - and they were able to explain how they ran another circuit from the same batteries that were running the motor. The boys were especially impressed.

During our institutes, students are always encouraged to share discoveries and teach one another. While working on chemical rockets one summer, students built their own double-safety ignition switches entirely from scratch. One student who had completed his project early started experimenting with clipping his switch's jumper leads to a copper penny and various other coins, then a metallic paper gum wrapper, then a soda can, a paper clip, the graphite from a pencil, and so on. He was making discoveries about the conductivity of different objects and materials. As the other kids started to notice what he was doing, he began explaining his findings and encouraging them to follow suit. The rest of the afternoon turned into an inquiry session on conductivity led by this self-motivated student.

Sometimes it takes a while for a student to relax and realise that being in an *Exploratorium* workshop is very different from being in a classroom, and that he or she isn't being graded. One such student, a boy named Justin, needed a lot of support from the staff to complete his first project, which was building an animation machine called a zoetrope. Justin was worried that he'd make mistakes or that his zoetrope wouldn't work as well as everyone else's. During the second and third weeks of the five-week institute, Justin gained confidence. By the fourth week he was fully engaged in building his final project (a motor car) and was experimenting freely with no fear of being judged. He had modified his car wheels, making the back ones much larger than the front ones, and then discovered that putting rubber bands around his wheels gave his car more traction and a better ride on both carpet and bare ground. Justin was elated by the fact that his rubber-band tire treads were then copied by many of the other students.

David, 11 years old, was very unfocused and had been having a difficult time at school. But the prospect of being able to build a zoetrope, design his own cartoons, learn how to use metal-working tools to build wind chimes, and make a motorized model car from scratch was his motivation to get on track. Working on his projects, he was able to transform his hyperactivity into constructive energy. By the third week of the institute, David had acquired the habit of paying close attention to instructions and explanations from the staff. He'd say, 'Show me again! Show me again!' until he felt comfortable with the information he was being given. David did a lot of experiments with his motor car - especially in trying to find the right number of batteries to make it go faster. Initially, most of the kids made the assumption that more batteries meant more power and a faster car. But David discovered that adding batteries also added weight, which resulted in more drag. Several of the other children followed his lead and spent time experimenting with more power/greater mass versus less power/smaller mass. When we eventually met David's mother, she was so happy to tell us that, for the first time in his life, David was excited about going to 'school'.

When Nora started her first summer institute at 12 years of age, she was argumentative, annoyed at having to participate in conversations about the science experiments she was doing, and far more interested in socialising than learning. As an adolescent with visual problems (Nora is legally blind), she was uncomfortable at school and, although quite smart, had been placed in a special-education class.

Learning how to work with tools and building two sound exhibits, one on resonance and another on focusing sound waves, helped her gain self-confidence and an interest in science. She enjoyed herself enough to enroll again the following summer. During her second five-week institute, Nora decided to challenge herself by choosing to be on a student team that built six visual-perception exhibits. She helped out younger students in the machine shop and led group activities and experiments. Nora also developed a wonderful ability to ask excellent questions that always instigated lively discussions.

2.3. Exploratorium *Teacher Institutes*

Even though we are talking about informal science outside the classroom, the student projects at the Science Community Workshops often have solid connections to the formal science children study in school. Even for teachers working in formal settings, there can be much to learn from an informal educational environment. When teachers are offered the same workshop learning experiences, their reactions are much the same as those of the children. In formal evaluation studies of our Teacher Institute program (Inverness Research Associates, 1997; Allen, 1997), researchers found that the teachers who built classroom-sized science exhibits in the *Exploratorium*'s workshops made conscious connections between designing and building those exhibits and understanding the related science concepts. Said one teacher:

> I had no idea of what resonance meant, so I hoped I'd understand it more as I built (resonance rings)...and I played with it, and yes, I think I understand resonance: that a motion can be transmitted from one object to another through its vibrations. (Allen, 1997)

Like their students, the teachers also seemed to take pride in their accomplishments and expressed a desire to share their experiences with others:

> I am so proud I could make something, something that's fascinating and tricky. What have I made before? Dishes of food, flower arrangements. But this is scientific. And the opening part of the (teacher) institute was images and mirrors, so this (exhibit I built) is a constant reminder of the institute, not a memento, but an illustration. (Allen, 1997)

Finally, this workshop learning did ultimately transfer back to the classroom (Kaufman and Allen, 1997) and even to home. One teacher enthusiastically reported that her husband now grows nervous when she comes home and heads toward the garage where his power tools are stored! The changes are significant.

2.4. *An Example of a Teacher Demonstration*

One of the activities we do with children in grades 4–8 is have them design their own apparatus to measure the calories found in different foods. Students begin by watching a walnut burn: most are amazed to find that a walnut can burn for up to seven minutes. After this demonstration, the teacher asks students to explain, in their own words, what a calorie is. Most students associate a calorie with fat, gaining weight, or something their parents are now counting; no one connects a calorie to a unit of energy. Students are then told that a calorie is the amount of heat needed to raise the temperature of one gram of water one degree Celsius. Once they have that information, we give them a pile of aluminum cans, some coat hangers, aluminum foil, thermometers, and tape - and then, with no instructions, we ask them to invent and build their own devices to measure the calories in different foods.

This approach is straight out of the experimentalist tradition. The pedagogy is very powerful because students must have a true understanding of the concept in order to design their devices. How each student approaches the experiment reveals much about his or her understanding of the question; design failures often lead to very fruitful questions. This experience also gives students a great sense of confidence, not just because they're designing an experiment themselves, but also because they are being trusted to work with fire. Most students are not allowed to do 'dangerous' activities - especially ones they design themselves.

2.5. Family Events

We deal with many children who are from diverse cultures and speak other languages at home. Having parents and children doing projects together is extremely helpful for all concerned. At a time when students are being acculturated faster then their parents, and sometimes growing distant because of this, these activities serve as bonding agents. Most cultures outside the United States tend to be family and community oriented, so family science events tend to bring families together. Many of the projects we do with families have a way of connecting new cultures with the old. For example, in one family workshop, we have the children make and use stilts. In Latin America, stilts are very common, so this fun activity serves to bring some families together. We also try to find activities that deal with art, which again crosses language and brings families together.

3. TRICKS OF THE TRADE

The stories above portray some of our best examples of success. We have had our failures; indeed, we have learned much through trial and error. Our rate of success is becoming more consistent, however, and we want to illuminate some design features that have increased our probability of success, so that they may serve others who attempt to replicate this workshop approach to learning.

3.1. Rules

We do not lay down an overwhelming number of explicit or pre-determined rules. The workshops have to be oriented toward the children, so when rules of conduct are established, the place and the children dictate what these rules should be. The founding students set the rules by example, and the new students respect and understand their importance, so implicit rules are automatically followed. Our sites have few problems with fights or arguments. Many children who typically give teachers problems are no problem in this type of place. This comes from a feeling of ownership.

3.2. Safety

The rules that are made in advance should be mainly about safety issues. Emphasise safety as an issue that everyone–both children *and* adults - must be aware of at all times. If safety rules are not seen as meaningless restrictions, but as guidelines to more successful building, the children will follow them.

Before anyone starts building anything, do safety training for each child on each and every tool or piece of machinery. Periodically review training. Have a safety zone marked out on the floor in colored tape around each piece of machinery so that children do not crowd each other while working. Make sure each child wears appropriate safety equipment: goggles, dust masks, and gloves. We never let any student work in the shop alone. There always has to be a responsible adult present and paying attention.

We usually work with middle school-aged children 11–14 years old; we do not recommend kids that under 9 or 10 work with shop tools and machinery. Using safe materials, like the cardboard, paper, scissors, tape, and glue required for building a paper zoetrope, cardboard spectroscope, or a simple magnet toy, can, however, be very rewarding for younger children. There is nothing equal to the feeling of being able to say, 'I made it myself!'

We have been very fortunate that in many thousands of hours in over a dozen locations, we have had no serious injuries. We typically have hot-glue burns and minor scrapes and cuts. We spend quality time teaching students how to keep themselves safe, teaching them, for instance, to wear gloves when using hot glue and safety glasses when using power tools. We're also not above putting a little scare in the kids. Accidents happen when you get complacent. A little fear is good.

3.3. Cheaper is Better

The more common and inexpensive the material used, the more it contributes to the demystification of science. It also makes recreating projects easier if, say, the child wants to build the same thing again at home or at school or help a friend build one. Try to make your place feel like a club that one of the students might have started.

Use familiar, everyday materials so students can continue to tinker and explore wherever they are. You do not need special tools or materials to do science.

3.4. Use Students as Teachers - and Avoid the Role Yourself

We have had great success with older children mentoring or assisting younger ones. Occasionally, children become frustrated when they think they can't do a project. Let the child struggle a bit to see if he or she can work it out. If you do help, help in increments. Don't give it all away at once. Let the children help one another. Children often learn better from their peers.

3.5. Be Aware of Gender and Ability Equity

Don't treat the girls any differently from the boys or assume that boys have either more skill or more affinity for working with tools, wires, and machinery than the girls do. Don't discriminate against children with disabilities. Try to devise ways so that everyone can complete a building task and feel good about it, even if they need assistance from staff members.

3.6. Space: Funky is Better

Too formal an atmosphere or too tidy a workshop can intimidate children by making it seem uninviting, too much like school. Make sure you have a wealth of materials available so that students can modify, improve upon, or play around with the original design of a project (safely, of course). Allow for an artistic element when appropriate, so that children can personalize their projects.

3.7. Shortcuts

Each step of a project can be articulated clearly in pictures or photos so that kids can work through several steps on their own without taxing staff members with constant questions about what to do next. Once in a while, we will have some parts of a project pre-fabricated. To build part of a zoetrope, for example, students each need twelve uniform pieces of masonite. In the interests of time, we'll have a staff member pre-cut these pieces, since the students have only a week and a half to complete their projects.

4. WHY THIS WORKS IN CROSS-CULTURAL SETTINGS

Not only does this 'build it yourself' approach seem to work for large numbers of students, but our observations suggest that it helps cross many of the same types of boundaries that vex educators in more formal settings. Although these are our informal conjectures about the workshop environment, it is our hope that others might formally research them.

(i).   It is an equal-opportunity environment. We trust everyone with the equipment and the materials. Where else does an 11-year-old get to handle a drill press, neighborhood 'roughs' get put in charge of building a mini-museum, or a group of grandmothers get access to power tools?

(ii).   It is not like school. This is not a criticism of schools; rather, it acknowledges that many of these students associate feelings of frustration and failure with school experiences. At the Workshop, students (and others) can start anew.

(iii).   It is a sex equalizer. The workshop is no longer a male domain. Many women are surprised to find how well they handle building tools.

(iv).   It is an ethnic equalizer. No racial or ethnic group has a primary claim on being the superior builder. Working with your hands and with materials is a universal instinct.

(vi).   It is not language dependent. Even if you cannot speak the primary language, you can still build something, show it, and convey meaning through the object.

(vii).   It engenders respect. The students have greater respect for the things they build and the places they are responsible for. The issue of ownership and pride mitigates behavioral problems and promotes sharing.

(viii). It ensures success. The students make the models work. Success is obvious and tangible when an exhibit or apparatus does what it is supposed to do. When something does not work, the problems are often obvious and adjustments easy to make. The result is empowerment and the student becomes an expert over at least that one thing.

(ix).   It promotes a goal: the goal is inquiry. Students are not penalized for failing to find a single 'right answer.'

(x).   It is a safe place for risk and failure. Both are rewarded. Only inactivity is unacceptable.

(xii).   It is a place for creativity and imagination. The students make things they can take home, show off, and continue to learn from.

CONCLUSION

In essence, what we are describing here are places of learning for all children, but most certainly for those who may not have found their niche in school. Workshop environments can help bring these children back into the science literacy game, sometimes with startling results. However, we do not want to suggest that these are the only places where this type of learning experience can occur. We sometimes think of informal science happening primarily in science centres where there are lots of interactive exhibits. Technically, informal learning environments encompass the many places where learning can occurs outside of formal settings like schools.

We should not forget that quality environments and experiences like these don't always need big buildings, budgets, and staffs. Indeed, perhaps we should celebrate the smaller and less organised places more.

Center for Teaching and Learning, Exploratorium, San Francisco

REFERENCES

Allen, S. (1997). *Formative evaluation of the Exploratorium Teacher Institute*. Unpublished evaluation report. San Francisco: Exploratorium.

Allen S., & Kaufman, J. (1997). *Teacher Institute formative evaluation: A classroom-based study of impact on recent alumni*. Unpublished evaluation report. San Francisco: Exploratorium.

Inverness Research Associates. (1997). *Inquiry at the Exploratorium: Highlights of the 1997 Teacher Institute*. Unpublished evaluation report. Inverness, CA:

Rhodes, R. (1986). *The making of the atomic bomb*. New York: Simon & Schuster.

EPILOGUE

REFLECTIONS OF AN EMINENT SCIENTIST

P.C. DOHERTY

18. LEARNING SCIENCE COMMUNICATION ON THE JOB

1. INTRODUCTION

Most professional scientists have no formal training in communicating science to the broader community. Exposure to the 'big time' of newspapers and television often comes very suddenly, and with little warning. High quality science journalists working for prestigious dailies like *The New York Times* may contact you because your latest 'breakthrough paper' is to be published in a leading format, such as *Science*, *Nature* or the *New England Journal of Medicine*. Your Institution may push you into the limelight because of a perception that your brilliant work on cancer may help to attract philanthropy. In my case, I became interesting when I shared the 1996 Nobel Prize for Physiology or Medicine with my Swiss friend, Rolf Zinkernagel. Being a Nobel Laureate confers an instant and identifiable personality status in a media world obsessed with personality.

When put to the test, some scientists are natural communicators while others find such interactions to be difficult. The successful ones are open to different perspectives, think quickly and are comfortable with unstructured situations. Even people with superior communication skills can benefit from an awareness of some basic rules. I am not sure that I have these rules right but I can provide a perspective gained from personal experience over the last four years. There are pitfalls, only some of which can be avoided.

2. THE MEDIA ARE INTERESTED IN GOOD STORIES

Modern journalism is largely about telling stories. Science journalists who work in major newspapers will have the sophistication to extract that story from your research paper and the interview that follows. This is why they contacted you in the first place. However, most journalists have little (if any) science education and are talking to you because you are involved in something that looks to be newsworthy. You must be patient when talking with them, and take the time to go over the same ground until the meaning is clear.

Interviewers will be looking for statements that intrigue their viewers and readers. The way we talk to each other as scientists at specialist meetings is often too beset with qualifications, and dry as dust for those with totally different backgounds. You need to put yourself into the perspective of someone who is browsing through a newspaper at breakfast, or watching TV after a hard day with the kids or at the office.

S.M. Stocklmayer et al. (eds.), Science Communication in Theory and Practice, 271–278.

The print journalist will be thinking both in terms of writing an article that people will want to read and whether it will ever get past his or her sub-editor and be seen by anybody. The TV interviewer will be concerned about the easy flow of what is being said. The talkback radio host wants the listener to be intrigued or enraged to the point where they call in. Different formats have different needs.

Your basic motives will not always fit with those of the reporter or the interviewer. In a sense, each is trying to use the other. Such is life! The more you can structure what you want to convey in a form that suits the media's rules of engagement, the more successful you will be as a communicator.

3. USE LANGUAGE LIKE A NORMAL HUMAN BEING

Scientists talk funny! I do not mean by this the specialist jargon that is peculiar to any research discipline. Where many scientists go wrong in the public arena is to use words like 'abrogate', 'quantify', 'activate', 'deduce', 'elucidate' and 'modulate'. The words are Standard English, but many people will never say them. Think in terms of speaking to your grandmother, or to Uncle Fred who drives a tractor-trailer. Do not assume that you are any smarter than Fred or Grannie (you share some of the same genes); it is just that your training is different and you may have had a better start in life. Fred may make a lot more money than you do!

The type of language that is perfect for a large scientific meeting is often a disaster with a public audience or an arts-educated journalist. I sometimes sit there listening to colleagues putting a reasoned, logical case that I find extremely boring to a lay audience and is obviously deadly dull for someone who is not trained to endure this type of monologue. If *I* am thinking 'beam me up, Scotty', what is the reaction of those who are less accustomed to being assaulted in this way? A tedious talk is a form of insensitivity! It is better to be over-stated and flamboyant than to be boring! Nobody is recording what you say for posterity.

It is obviously important to avoid specialist terms, or to use them very sparingly and take the time to say exactly what they mean. There are some exceptions to this rule: AIDS, for example, is now part of the common language. Otherwise, it is usually possible to avoid using the S-word completely, with S standing (of course) for science. One thing that becomes apparent very quickly when you take the trouble to address the terminology problem is that we do not always think through what is meant by a particular descriptor. Scientists can be as woolly in their thought processes as anyone else, and are often less intellectually rigorous than, for example, a well trained lawyer. A place to start the clarification process is with your graduate students and fellows. Go through their writing efforts and systematically take out as much jargon as possible. Insist that they attempt plain English, not science-speak when they give a seminar or present at a laboratory meeting. This will be a therapeutic exercise for you and for them. It will help build everyone's communication skills.

4. THE CENTRAL ASSUMPTIONS OF SCIENCE ARE NOT SHARED BY THE BROADER COMMUNITY

Always be aware that a good deal of reporting about the consequences of science relies on the fact that most people are not trained to think in terms of probability and relative risk. A classic case is where little Charlie suddenly develops condition 'x' after receiving a vaccination. The fact that millions of kids get the vaccine makes such occurrences an absolute certainty. This seems almost impossible to get across and is, in any case, standard fodder for both print and broadcast journalism in bad news - that is, calm - periods.

A similar situation occurs with Dr Y, when people who believe that they benefit from Y's treatments are denied access to Y's wonder drug or public funding to pay for that drug. The fact that nobody has done a double-blind trial and Y is unwilling (or unable) to document its beneficial effect to the satisfaction of the regulatory authorities is irrelevant. The officials and volunteer members of the professional community who are charged with protecting the public interest are easy media targets. Responsible scientists are presented as being arrogant, narrow and self-serving, while civil servants are uniformly shown to be heartless, rigid bureaucrats.

Such stories are stock in trade, especially for the foot-in-the-door, knowing sneer type of reporting that is popular with a few members of the visual media. Of course, if Y's drug is licensed prematurely and a disaster follows, the same people will make hay with this and take no responsibility for any pressure that they might have applied. As Winston Churchill said of one of the British press barons, 'Power without responsibility, the prerogative of the harlot throughout the ages'. The best way to deal with the authoritarian-stooge type of set-up is to make sure that the people who talk to the journalists are both emotionally tough and good communicators. They need to come across as attractive, sympathetic and flexible, especially on TV. This can help to subvert even the most aggressive and biased interview.

Anyone who is nominated by an Institution as a media representative should first have a screen test! The worst possible scenario is to give the job to good old X, perhaps because X's job description includes talking to the press or because his active research career is over and nobody quite knows what to do with him. Find someone who is smart, personable and informed to do this very important job! If the Institution is involved in an area that is controversial, it is also very important that the contact person should be prepared for the fact that some interactions with the press will inevitably be disastrous. An individual who becomes hostile after taking a few hits should be removed quickly from this role.

5. AVOID THE NEGATIVE

Journalists are attracted to negative statements for several reasons. The first is that the general public are suspicious of people who put themselves forward (or are put forward) as specialists and are very happy to hear that their poor opinion is justified. The second is that a negative article often leads to follow-up stories as those who are criticised seek to defend themselves. Controversy itself is newsworthy, and can be more interesting than the subject that is in contention.

Scientists often fail to appreciate that people who are in the public eye, particularly professional politicians, will not react well to stories that are highly critical of them or their policies. If the aim is to achieve an increase in government funding, it is generally much more productive to emphasise the positive contributions that research and science education make than to demonise the Minister of Science for not recognising the obvious.

In fact, the politician responsible for the particular area is probably trying to secure a bigger share of tax-payer's dollars. An attack from the professional constituency can weaken his or her position with the political leadership. On the other hand, well-documented examples of the public benefits of research can provide ammunition for future arguments. Remember that 90% of the stories that any politician reads about him or her self will be negative. These people work hard and are generally dedicated public servants. Where possible, it does not hurt to be nice to them and to give them something to work with. Even if they are, for some reason, hostile, making the effort to establish dialogue and discussion will generally work better than confrontation. A process of mutual education can be very beneficial for both elected and appointed public servants, and for those from the scientific community.

6. IT DOESN'T MATTER: BE PREPARED TO TAKE COLLATERAL DAMAGE

The only safe media format is unedited radio or television. If you make a mistake, it is your mistake. The print media can do absolutely anything that they want with what you say. There are various factors operating in this. The interviewer may actually like you and what you stand for. The result will be a generally positive account. A journalist who is thinking in terms of talking to you again will try to be fairly accurate, so you are usually on reasonably safe ground with your local media outlets. However, you can be given a sense of false security if such positive interactions are the total extent of your experience. Anything can happen when you go on the road.

The culture of some newspapers is to trivialise almost everything. The journalist knows what the sub-editor will accept, and will seize on the aside you make, at the end of what you might have regarded as a serious interview, to focus on the human interest angle. With repeat performances, you get to learn how a particular paper operates and thus can exercise some pre-emptive damage control. My personal sense is that many editors greatly underestimate the breadth of interest of their readership–but I do not have to make my living selling newspapers.

Apart from whatever the particular journalist writes, you are totally at the mercy of the sub-editor. Another story line for the day may be that university academics are upset about the latest spectrum of funding cuts, or at the efforts of a university administration to move-out unproductive faculty. If this is the case you may find that what you thought was a positive speech about, for example, opportunities for the financial community in the application of molecular medicine, will appear the next day in the paper with some sort of hostile slant. The reason is that you got stuck on page 3 with the unhappy professors and, for consistency, the sub-editor ensured that whatever was reported about your talk was cut to the same negative cloth. This happened to me with one of the major Australian dailies, but was a single incident in what has generally been a good experience with that paper.

The important lessons to take away from a negative experience are that no matter how careful you are, there is nothing that you can do to control what appears in a newspaper; and it doesn't matter. Most newspaper stories are ephemeral, and have no legs. It isn't like something in a bound scientific journal that can damn you in perpetuity with your professional colleagues.

7. NOTHING YOU SAY IS AS INTERESTING AS PLUGGER LOCKETT'S GROIN

Plugger is a well-known and colourful Australian footballer. His pelvic injury was consistently page 1 stuff in the major Sydney dailies for weeks. Any reporting of science was lucky to make it to page 5. Albert Einstein would probably be on the front page if he reappeared on earth again, but Albert is an icon.

Major league sport provides a spectator activity that gives a sense of excitement, community and meaning to many people. The practice of science tends to be introspective and inaccessible, though the consequences for society can be enormous. Many journalists seem to expect that, as a scientist, I will be angry about the massive attention given to sport. In fact, it is worth emphasising that the attitudes of scientists to sport pretty much mirror the spectrum for the broader community. Some scientists relax by watching sport. A number are aggressive participants. Being a player is very much in character, as science itself is, obviously, for doers rather than observers.

It never hurts to make the point that, outside their obscure obsession, scientists are just folks. Scientists climb mountains, play football and tennis, and watch the Superbowl on television. One of the things that we have to get across to young people is that being absorbed by the excitement of science does not mean that you must be submerged in an excessively focused, nerd culture. It doesn't hurt for high school students to know that junior scientists sometimes drink too much beer, have great parties and get to ski powder in the Rockies. We should make more effort to put some of the more colourful, young science personalities in front of the media!

8. ALL MEDIA OUTLETS ARE IMPORTANT

We tend to think we have achieved a breakthrough if we get a science story published in one of the prestige dailies. This is great, of course, but the people who take the trouble to read the article are likely to be well educated and positively inclined towards science. When it comes to conveying a sense of the excitement and power of contemporary science to a much broader public, however, the most powerful organs can be outlets like talk-back radio or local, community newspapers. Many no longer buy a daily paper - it is the ingrained habit of the pre-TV generation. They do, however, read the throw-away weeklies. The journalists who work for these outlets don't seem to me to be any less sophisticated than those employed by the major formats, though they are often younger. I have had the experience of being interviewed on a talk back program, then listening to people who call in to say that they would greatly appreciate more science stories. The talk back people, in particular, are always looking for interesting material to fill their air time.

Those who listen to talk back programs as they drive to work may never tune the TV to a popular science program on the public network. An exception to this rule seems to be with stories about the environment (or animals in the wild) on formats like the 'Discovery' channel. The word 'ecology' was almost a specialist scientific term when I first went to university. Now it is part of the general culture. The keen, often well-informed interest in environmental issues is a good example of the power that TV has to shape public attitudes in a science-based area. It also tells us that people are willing and able to absorb such arguments when they relate to topics of general concern.

9. COMMUNICATING WITH THE POWER BROKERS

Most of the funding for basic science comes from the tax base. An exception is biomedical research in the United Kingdom, where the Wellcome Foundation now dispenses more money than the Medical Research Council. Wellcome is, of course, led and administered by people who have been (or are) professional scientists.

The ranks of professional politicians contain few scientists. Parliamentary systems that select their ministers from the body of elected representatives will often give the responsibility for science to people who come from very different cultural backgrounds. Fortunately, most such people are smart and keen to learn, though they do tend to be incredibly busy. The science portfolio may also be very minor or (as in Australia) different areas of science may be divided between several ministries. Health traditionally does relatively well in the Australian system, both because medical research is popular with the public and because health care budgets are so vast that the research component is quite small. However, much basic science comes under education. I have yet to see any reasoned explanation for, or positive benefit from, this arrangement.

Those scientists who live in Britain or Australia may well look with envy at their colleagues in Taiwan, France or the United States, where cabinet members with responsibility for science are appointed directly from the senior scientific community. The counterbalance to this approach in a parliamentary system should be a well-motivated bureaucracy that is at least positively disposed towards science and is able to present lucid arguments to the responsible minister.

The more open political system in the USA makes it very worthwhile to communicate with, and lobby, individual senators and congressional representatives. Big increases in medical research funding were achieved by Congressman Porter and Senator Hatfield, republican members of the House and Senate budget committees who saw the support of the National Institutes of Health as a major, personal priority. Both built tremendous reputations as a consequence, and were honored by the very prestigious Lasker Public Service Award.

I started out by urging Australian scientists to do the same thing. Get to know your Senator or House Member! This can have no negative effect, and you may be establishing a connection with someone who will later be given important responsibilities. However, the decision-making process in the British-type parliamentary model is restricted to the cabinet room and is neither transparent nor, so far as I can determine, much influenced by input from the back-benches. It is very important in this context for prominent members of the scientific community to estabish amiable contact with the particular cabinet member and his/her senior civil servants. Also, organisations like Scientific Academies need to work very hard at establishing a reputation for giving good advice, which in turn requires that those with real expertise provide their services wholeheartedly (and in a judicious way) when asked to serve.

10. IN CONCLUSION

The era when scientists (or any professionals) were accorded instant respect is long gone. Like every other group of people with specialised knowledge and insights, scientists must be prepared to interact positively and openly with the general public, the media and political decision makers. Sometimes this is a very difficult balancing act. Many journalists are much more interested in emphasising the negative in whatever you might say about public policy, which in turn has the potential to anger politicians and bureaucrats who may well be trying very hard to do the right thing. The politicians themselves realise this, as such experiences comprise so much of their public life, but it helps a lot if they know you personally and realise that you are not operating from motives that are destructive or related to party politics.

Like biographies, newspaper articles and television interviews tell us as much about the reporter as they do about the particular subject. You must be prepared for the fact that you will, at times, be used to push agendas other than your own and accept that reality with good grace. It is impossible to win a fight with a newspaper or a TV channel but, if you make the point gently that you were not thrilled about what happened, you will probably be treated with more respect the next time around.

I have not said much in this chapter about speaking directly to the general public. Though the media can act as a multiplier, particularly if an address is recorded and then re-broadcast several times, the personal experience of hearing and meeting a well-known and interesting speaker can have an enormously positive effect. The rules are fairly obvious. You should be simple, clear and should not talk down to the audience. It is essential to realise that practicing research scientists start from a background and set of assumptions that are not shared by most people, including many who have professional training in science-based areas (eg pharmacy, dentistry). Treat those who ask questions with respect, even if what they say seems to you to be obtuse or perhaps hostile. You must accept that some people may simply dislike whatever it is that you are trying to convey. It doesn't help to antagonise them further, especially if the confrontation makes you look boorish or remote.

Above all, stay cool, lucid and retain your sense of the absurdity of human pretensions, especially your own. If you are by nature stiff and unable to laugh at yourself, stay off the public speaker circuit and restrict your presentations to set pieces, the written word or meetings attended by your long-suffering peers. The aim is not to dictate, nor to be didactic. You are not is standing up there as some sort of superior class-room teacher. This is one of the many reasons that I dislike the term 'professor'. Just as bad is 'doctor', which has the same effect of establishing a spurious authority that distances the speaker from the audience. Try to get those who introduce you to call you 'Bill or Betty Jones'. The need is to communicate, to establish a personal and sympathetic connection with those who (from whatever motive) have taken the trouble to come to hear you. Remember always that your basic aim as a public speaker is to provide people with something new and exciting to think about. They have given you their time, so you owe them something worthwhile. The task is as much to convey an enthusiasm for science and rational enquiry as it is to inform. The product is a good one, but it does not always sell itself! You may even discover that you enjoy being an entertainer and a salesman of ideas.

Department of Immunology, St Jude Children's Research Hospital, 332 North Lauderdale, Memphis Tennessee 38105

NOTES ON CONTRIBUTORS

GLEN AIKENHEAD

Glen Aikenhead is Professor of Curriculum Studies at the University of Saskatchewan. He has always embraced a humanistic perspective on science, even as a science teacher at international schools in Germany and Switzerland. This humanistic perspective was enhanced during his graduate studies at Harvard University, and has since then guided his research and scholarly writing related to: curriculum policy, student assessment, classroom materials and instruction, and cross-cultural science education. This research and development has resulted in such books and documents as: Science: A Way of Knowing (1975); Science in Social Issues: Implications for Teaching (1980); Views on Science Technology and Society (VOSTS, 1989); Logical Reasoning in Science & Technology (LoRST, 1991); STS Education: International Perspectives on Reform(co-edited with Joan Solomon of Oxford University, 1994); and Rekindling Traditions: Cross- Cultural Science & Technology Units (2000).

IAN ALLEN

Before becoming involved with multimedia and the World Wide Web, Ian Allen spent 10 years as a TV science producer for the Australian Broadcasting Corporation. In 1996 he wrote and directed the award-winning science CD-Rom *Ingenious*. Since 1997 he has been the producer of the ABC's popular Online Science gateway – *The Lab*.

VIVIAN ALTMANN

Vivian Altmann started the Children's Educational Outreach Program at the Exploratorium in 1985 and directs it today. The Outreach program partners with neighborhood centers and community organizations to provide them with a wide range of educational services for children and families. Specifically, this program develops curriculum, teaches workshops, facilitates professional development sessions, coordinates family events, and participates in the research and development of Exploratorium publications and products. Ms. Altmann also serves as a consultant to science educators all over the world who are starting educational outreach programs in museums, universities and other settings. She has worked at the Exploratorium for more than 20 years.

DENNIS BARTELS

Dennis Bartels is director of the Center for Teaching and Learning at the Exploratorium. The Center is designed to use the exhibit and staff resources of the Exploratorium to facilitate science education improvement efforts locally and nationally. Before joining the Exploratorium, Dr. Bartels was principal investigator and project director of the National Science Foundation sponsored South Carolina Statewide Systemic Initiative and directed the development of the State Curriculum Frameworks there. He is chair to the State Advisory Committee of the California Science Project and serves on several committees, advisory boards and review panels for the National Science Foundation and other education organizations.

PETER BRIGGS

Peter Briggs was born and brought up in Sheffield, England. He studied chemistry and completed a doctorate in theoretical chemistry at the University of Sussex in Brighton. After two post-doctoral posts, he spent seven years working in non-formal education within the UK on behalf of overseas-aid agencies in the voluntary sector. In 1980 he joined the BA where he was responsible for many different aspects of its work before becoming Chief Executive in 1990. Peter has extensive experience of organising science communication programs and activities. He is involved in a variety of UK and international initiatives and committees concerned with science communication.

CHRIS BRYANT

Chris Bryant is an Emeritus Professor at the Australian National University and a Visiting Fellow at the National Centre for the Public Awareness of Science. Australian University. He was its first Director. A graduate in Zoology and Biochemistry from London University, his research on the host/parasite relationship is described in more than a hundred scientific papers and several books. In 1963, he moved to the ANU, taking up a lectureship in the Zoology Department, eventually being appointed to the Chair of Zoology and subsequently, to the position of Dean of Science. He retired early, at the end of 1994, to concentrate on science communication and the Centre, where he continues to work. He is a Fellow of the Institute of Biology (UK), the Australian Institute of Biology and the Australian Society for Parasitology. In 2000 he was made a Member of the Order of Australia (AM) for services to science communication, research and education.

DEBORAH COHEN

Deborah Cohen has worked in science broadcasting since 1979 when she joined the BBC Radio Science Unit as a researcher. Through the 1980s she worked producing radio programmes for Radios 3 and 4 on a wide range of subjects in science, technology and medicine. She worked briefly in TV and then became Editor Science for domestic radio output in 1990. In October 2000 she also took on editorial responsibility for the Science Unit of BBC World Service. Deborah has a degree in Physics from the University of Durham, and a M.Sc. in Social Studies in Science from the University of Manchester. She was a judge of the Rhone Poulenc Book

Prize in 1997; has been a member of the COPUS committee that awards grants for the public understanding of science; and regularly gives talks about the presentation of science on radio and TV.

GOÉRY DELACÔTE

Goéry Delacôte is Executive Director of the Exploratorium, San Francisco. He is also Professor of Physics, currently on leave, from the University of Paris. From 1982 to 1991, Dr. Delacôte was Director of Science and Technology Information, one of eight scientific divisions of the Centre National de la Recherche Scientifique (CNRS). His field of interest is scientific education and information.

PETER DOHERTY

Peter Doherty AC, FRS, FAA was awarded the Nobel Prize for Physiology and Medicine in 1996. A graduate of the University of Queensland, he completed his PhD at the University of Edinburgh in 1970. After a spell as Veterinary Officer at the Animal Research Institute, Brisbane, Australia, he returned to Scotland and a post as Scientific Officer at Moredun, Edinburgh. From 1972 to 1975 he was a Research Fellow at the John Curtin School of Medical Research at the Australian National University. It was during this time that he completed the work, with Rolf Zinkernagel,, that led to the award of the Nobel Prize almost twenty years later. There followed periods as Professor at the Wistar Institute in Philadelphia and Professor and Head of the Department of Experimental Pathology at the John Curtin School of Medical Research. In 1988 he returned to the USA, to the Department of Immunology, St. Jude Children's Research Hospital, and as Adjunct Professor in the Departments of Pathology and Pediatrics, University of Tennessee, in Memphis.

RICHARD ECKERSLEY

Richard Eckersley is currently a fellow at the National Centre for Epidemiology and Population Health at the Australian National University, Canberra, where he is researching aspects of progress and well-being. He is a former science writer for The Sydney Morning Herald and a former head of the Media Liaison Group at CSIRO, Australia's national research organisation. He has also worked for the (now defunct) Australian Commission for the Future and on the personal staff of a Federal Government Minister. He still writes occasional pieces for leading newspapers and the ABC and is frequently interviewed in the media about his work.

SIMON GAGE

Simon Gage trained as a physicist at the Universities of Bristol, Dundee and Edinburgh receiving his PhD in 1989 from Edinburgh. He joined the Edinburgh International Science Festival in 1990, set up the schools touring program in 1991 and then ran both this and the interactive family program until he became director in 1995. He has written popular science books for children and enjoys doing large scale public experiments and demonstrations including walking on hot coals.

JOHN GILBERT

John Gilbert is Professor of Education at The University of Reading, U.K. His initial qualifications were in chemistry and he taught that subject in high schools before returning to university work. For may years he was associated with research into students 'alternative conceptions' of the ideas of science and with the implications of these for teaching and teacher education. In recent years his research has focused on the role of models in science and technology education and on the part that these play in the provision of explanations. This work has extended beyond the school and the formal curriculum to interactive science and technology centres and to informal learning. For several years he has been associated with NCPAS and has worked with Sue Stocklmayer on visitors use of the exhibits at Questacon: the present chapter is based on that work. He is also Editor-in-Chief of the International Journal of Science Education and is encouraging the submission of articles on informal learning and on science communication to that journal

MIKE GORE

Mike Gore trained at Leeds University in electrical engineering and holds Michael Faraday as one of his idols. Mike is a native of Bolton, Lancashire and came to Australia in 1962. In that year he joined the academic staff of the Australian National University where he was to teach Physics for the next 25 years.
In 1980 his great love of teaching - both students and the general public - was the spur that led to him to establish, under the auspices of the ANU, Australia's first interactive science centre, *Questacon*. In the following decade he was for a time scientific adviser to the ABC television series "Towards 2000". In 1982 he was 'Canberran of the Year' and was awarded a Churchill Fellowship that enabled him to visit and study overseas science centres in North America and Western Europe. In 1986 he was created a Member of the Order of Australia (AM), for "his services to science education". The following year he received a special award from the "Beyond 2000" television team for his role in promoting public science education.
In 1987 he left academia and became the foundation director of *Questacon - The National Science and Technology Centre*. In 1992 Dr Gore and *Questacon* were jointly awarded the ABC's prestigious Eureka Prize for "the public promotion of

science". Mike retired as director of *Questacon* at the end of 1999 and returned to the ANU where he is now based in the National Centre for the Public Awareness of Science. He is an Adjunct Professor in Science Communication at ANU.

LAWRENCE J. PRELLI

Lawrence J. Prelli is Associate Professor of Communication and Adjunct Associate Professor of Natural Resources at the University of New Hampshire. His research and teaching interests are in rhetorical theory and criticism, with special emphases on rhetorical studies of scientific and environmental discourses. In addition to essays in scholarly journals and chapters in anthologies, he has edited two special issues of international journals on the theme of rhetoric and science and is author of *A Rhetoric of Science: Inventing Scienctific Discourse*, which won the Eastern Communication Association's Everett Lee Hunt Scholarship Award. He has served on the editorial boards of scholarly journals such as *The Quarterly Journal of Speech* and *Communication Quarterly*. Currently, he is editing a book on the rhetorical dimensions of display.

LÉONIE RENNIE

Professor Léonie J. Rennie is Dean of Graduate Studies at the Curtin University of Technology in Western Australia. She has a background in science teaching in Western Australian schools and was involved in teacher education programs at the University of Western Australia before taking up her position at Curtin University in 1988. She is interested in the ways in which students learn science and technology, in both formal and informal settings, their attitudes about science and technology, and gender-inclusive assessment of cognitive and affective learning. She has published widely in science and technology education and serves on the Editorial Boards of the *Journal of Research in Science Teaching* and the *Australian Science Teachers Journal*.

PETER SPINKS

Peter Spinks holds a Master's degree in research psychology and has published in international academic journals. Since 1980, he has broadcast and written for leading media organisations including the British Broadcasting Corporation, *The Guardian* (London) and *The Observer* (London) newspapers and *New Scientist* magazine. He is currently a senior staff journalist on a major daily newspaper in Australia. *Wizards of Oz*, his book about recent breakthroughs by Australian scientists, has been twice the No.1 bestseller on *New Scientist's* list of top-selling science books. Peter also teaches science writing and media skills at comprehensive workshops which he designs to meet the specific needs of universities, research organisations and companies around the world. Email: pspinks@hotmail.com

MODESTO TOMEZ

Modesto Tomez is coordinator for the mentors in the Teacher Institute at the Exploratorium. He is part of a team responsible for developing a new teacher induction program, which prepares experienced teachers to support first and second year middle and high school science teachers. He spent almost twenty years teaching K-12 in Spanish and English in Chicago, San Francisco, and San Jose. In the past ten years he has been involved in writing a set of multicultural math and science books and establishing science after school children centers throughout the state of California. He has also lead teacher workshops and taught in classrooms in Brazil, Chile, Mexico, Spain and Portugal.

RUTH SHRENSKY

Dr Ruth Shrensky, whose background is in philosophy, is currently a lecturer in study skills at the University of Canberra. Her specialities include philosophy of communication (her PhD thesis was a critical reappraisal of communication models, especially those used for public communication campaigns), and the teaching of academic research and writing skills.

DAVID SLESS

Professor David Sless is Director of the Communication Research Institute of Australia and Senior Visiting Research Fellow at Coventry University. His specialities are the philosophy of communication and information design. He is a frequently invited speaker at international conferences in North America, Europe and Asia, and is the author of over 180 publications.

JON TURNEY

Jon Turney trained as a scientist and historian of science, and worked as a science journalist (still does occasionally). Now his day job is head of the Department of Science and Technology Studies at University College London where, among other things, he teaches science writing to undergraduates. This started out on the not very persuasive premise that if you can do it, you can teach it. Since then, he has been trying to think about what differentiates science writing - and the problems science writers have - from other kinds of expository writing. He is also interested in developing a critical vocabulary for popular science. He author of Frankenstein's Footsteps: Science, Genetics and Popular Culture (Yale University Press, 1998), and co-editor with Maurice Riordan of A Quark for Mister Mark: 101 Poems about science (Faber, 2000). His next book will be a study of the popular science literature.

Science & Technology Education Library

Series editor: Ken Tobin, *University of Pennsylvania, Philadelphia, USA*

Publications

1. W.-M. Roth: *Authentic School Science.* Knowing and Learning in Open-Inquiry Science Laboratories. 1995   ISBN 0-7923-3088-9; Pb: 0-7923-3307-1
2. L.H. Parker, L.J. Rennie and B.J. Fraser (eds.): *Gender, Science and Mathematics.* Shortening the Shadow. 1996   ISBN 0-7923-3535-X; Pb: 0-7923-3582-1
3. W.-M. Roth: *Designing Communities.* 1997
           ISBN 0-7923-4703-X; Pb: 0-7923-4704-8
4. W.W. Cobern (ed.): *Socio-Cultural Perspectives on Science Education.* An International Dialogue. 1998   ISBN 0-7923-4987-3; Pb: 0-7923-4988-1
5. W.F. McComas (ed.): *The Nature of Science in Science Education.* Rationales and Strategies. 1998   ISBN 0-7923-5080-4
6. J. Gess-Newsome and N.C. Lederman (eds.): *Examining Pedagogical Content Knowledge.* The Construct and its Implications for Science Education. 1999
           ISBN 0-7923-5903-8
7. J. Wallace and W. Louden: *Teacher's Learning.* Stories of Science Education. 2000
           ISBN 0-7923-6259-4; Pb: 0-7923-6260-8
8. D. Shorrocks-Taylor and E.W. Jenkins (eds.): *Learning from Others.* International Comparisons in Education. 2000   ISBN 0-7923-6343-4
9. W.W. Cobern: *Everyday Thoughts about Nature.* A Worldview Investigation of Important Concepts Students Use to Make Sense of Nature with Specific Attention to Science. 2000   ISBN 0-7923-6344-2; Pb: 0-7923-6345-0
10. S.K. Abell (ed.): *Science Teacher Education.* An International Perspective. 2000
           ISBN 0-7923-6455-4
11. K.M. Fisher, J.H. Wandersee and D.E. Moody: *Mapping Biology Knowledge.* 2000
           ISBN 0-7923-6575-5
12. B. Bell and B. Cowie: *Formative Assessment and Science Education.* 2001
           ISBN 0-7923-6768-5; Pb: 0-7923-6769-3
13. D.R. Lavoie and W.-M. Roth (eds.): *Models of Science Teacher Preparation.* Theory into Practice. 2001   ISBN 0-7923-7129-1
14. S.M. Stocklmayer, M.M. Gore and C. Bryant (eds.): *Science Communication in Theory and Practice.* 2001   ISBN 1-4020-0130-4; Pb: 1-4020-0131-2

KLUWER ACADEMIC PUBLISHERS – DORDRECHT / BOSTON / LONDON

Lightning Source UK Ltd.
Milton Keynes UK
UKOW06f1815140715

9 781402 001314